Ian POPPLE Otto WAGNER Peter McDOUGALL

SCUBA DIVE SNORKEL SURF

PALM BEACH

FLORIDA

Published by Mango Publishing Group, a division of Mango Media Inc.

Series concept and writer: Ian Popple
Art concept and illustrations: Otto Wagner
Writer: Peter McDougall
3D models: Emil Stezar
Editing: Peter McDougall, Kris Garland

Front cover photo Rostislav Ageev/Shutterstock©, Warren Metcalf/Shutterstock©
Back cover photo patuletail/Shutterstock©

For permission requests, please contact the publisher at:
Mango Publishing Group
2850 Douglas Road, 3rd Floor
Coral Gables, FL 33134 USA
info@mango.bz

For special orders, quantity sales, course adoptions and corporate sales, please email the publisher at sales@mango.bz. For trade and wholesale sales, please contact Ingram Publisher Services at customer.service@ingramcontent.com or +1.800.509.4887.

Reef Smart Guides Florida: Palm Beach: Scuba Dive. Snorkel. Surf.
ISBN: (print) 978-1-64250-240-4, (ebook) 978-1-64250-241-1

Printed in the United States of America

Acknowledgments

Reef Smart is indebted to numerous individuals and organizations who contributed their advice, knowledge, artwork and support in the production of this guide. We would particularly like to thank Discover The Palm Beaches, whose financial support helped make this project possible. We would also like to thank Jena McNeal, with Palm Beach County's Department of Environmental Resources Management, for helping get the project off the ground. Thanks also to Shana and Dean Phelan at Pura Vida Divers, Adam Birdwell at Starfish Scuba and Gerry Carroll at Jupiter Dive Center for their help visiting the dive and snorkel sites featured in this guide, for sharing their knowledge of these sites and for their feedback. Thanks to Zach Nolan and Andrea Whitaker for the use of their photographs. Thanks also to Pavan Arilton at Dixie Divers and Bradley Williams for help with the joint Broward/Palm Beach sites. Additional support provided by Rick Netzel and the Palm Beach Lakes Best Western, the Waterstone Resort & Marina Boca Raton by Hilton, and the Palm Beach Marriott Singer Island Beach Resort and Spa.

Financial support provided by:

About Reef Smart:

Reef Smart creates detailed guides of the marine environment, particularly coral reefs and shipwrecks, for recreational divers, snorkelers and surfers. Our products are available as printed guidebooks, waterproof cards, wall posters, dive briefing charts, beach signage and 3D interactive maps, which can be used on websites and as apps. Reef Smart also provides additional services to resorts that are dedicated to offering an environmentally aware experience for their guests; these include marine biology training for dive professionals and resort staff, implementation of coral reef monitoring and restoration programs, and the development of sustainable use practices that reduce the impact of operations on the natural environment.

www.reefsmartguides.com

TABLE OF CONTENTS

How to use this book

Objective

The main objective of this guide is to provide a resource for people, particularly divers, snorkelers and surfers, who are interested in exploring the marine environment of Palm Beach County. This guide is designed to be used alongside Reef Smart handheld waterproof cards, which can be taken into the water. This guide will be most useful for watersports enthusiasts but also includes information that any visitor to southeast Florida will find useful.

Mapping

We have attempted to catalog the region's dive and snorkel sites as well as its beaches and surfing spots. However, only what we consider to be the top dive and snorkel sites in Palm Beach have been featured in more detail using Reef Smart's unique 3D-mapping technology. These maps provide useful information such as depths, currents, waves, suggested routes, potential hazards, unique structures and species information, which cannot be found in other guides. The maps in this guidebook are listed from north to south.

Disclaimer

Reef Smart guides are for recreational use only – they are not navigational charts and should not be used as such. We have attempted to provide accurate and up-to-date information for each site, as well as activities to enjoy in the surrounding areas. However, businesses close and new ones open, prices are adjusted and change is inevitable in the marine environment. The information contained in this guide is accurate only at the time of publication. The size and location of structures may vary. Depths and distances are approximated in both metric and imperial units, and the suggested route is optional. Reef Smart assumes no responsibility for inaccuracies and omissions, and assumes no liability for the use of these maps. If you do identify information that should be updated, please let us know at **info@reef-smart.com**.

Information boxes

Additional information for the featured sites is provided in the form of special information boxes, which appear throughout the book:

DID YOU KNOW? 
Interesting facts about the site or the surrounding area.

SAFETY TIP 
Advice that aims to improve safety.

ECO TIP 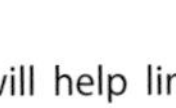
Information that will help limit damage to the ecosystem or improve environmental awareness.

RELAX & RECHARGE 
Information on where refreshments can be purchased, or where to unwind on land.

SCIENTIFIC INSIGHT 
Information of a scientific nature that can help you understand what you see and experience.

Map icons

SCUBA dive

Snorkel

Wreck

Access by boat

Access by swim

Access by car

Access by walk

Surf

Kiteboard

Wind surf

ECO TIP 

We hope this guide enhances your in-water experience. Share your passion for exploring the marine environment with others, because our oceans, and particularly coral reefs, need all the "likes" they can get. Coral reefs, as well as mangrove and seagrass ecosystems, are under serious pressure from a multitude of threats that include coastal development, pollution, over-fishing and global climate change. Some estimates put over half the world's remaining coral reefs at significant risk of being lost in the next 25 years; raising awareness can help protect them.

Species identification

The species listed for each location were chosen to represent the most unique or common organisms found at each site, as determined from personal observations, discussions with divers and snorkelers who have experienced these sites, and from scientific studies conducted in these areas. Many of the species described in this publication are mobile or cryptic (or both), and so may not always be found where indicated. However, we have attempted to place key species on each map in the locations where they are most commonly found.

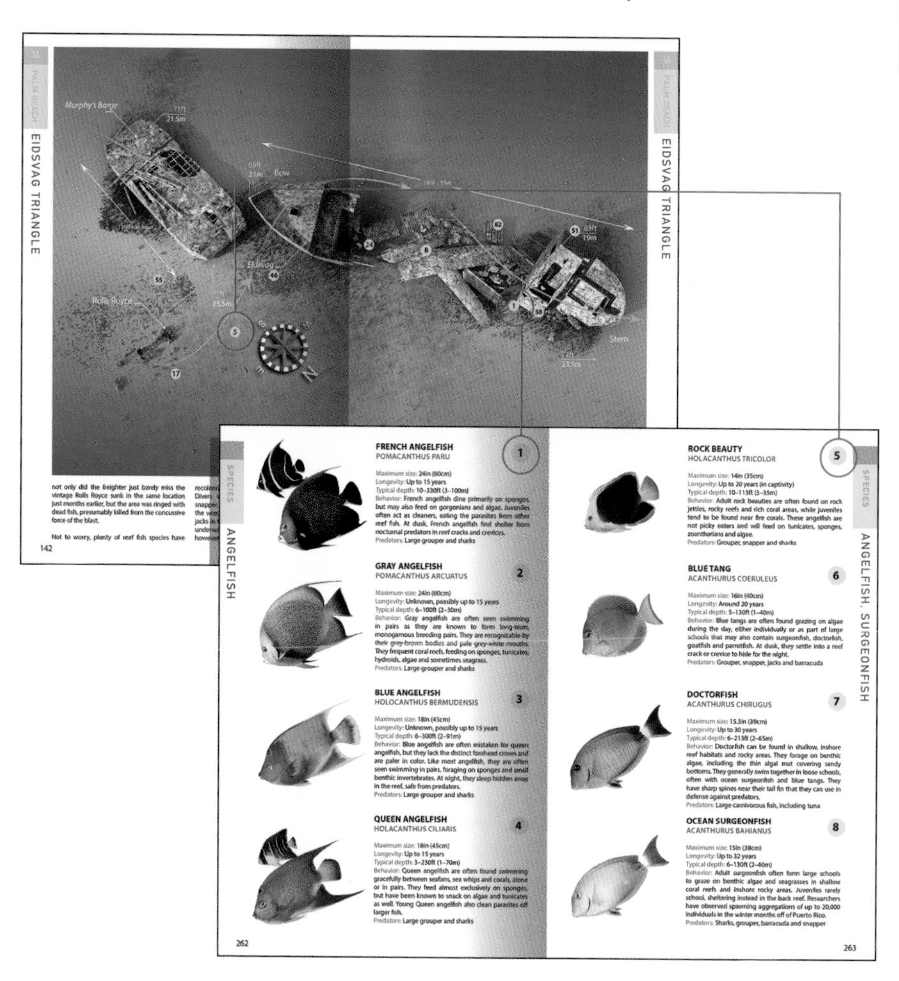

Species description

The species letters and numbers on each map link to descriptions located at the back of the book (on pages 252–284). Reef Smart uses the most frequently cited common name for a species. As common names vary from place to place, we have also provided the scientific name for each species, which remains the same worldwide. Scientific names are usually of Latin or Greek origin and consist of two words: a genus name followed by a species name. By definition, a species is a group of organisms that can reproduce together such that it results in fertile offspring; a genus is a group of closely related species.

The descriptions of each species are based on the scientific literature as it existed at the time of publication. Scientific knowledge often advances, however, and the authors welcome any information that helps improve or correct future editions of this guidebook. In-depth species profiles, including images and videos, are available for free on our website – **Reefsmartguides.com**.

Our "blue planet"

Oceans

Water covers nearly three quarters of our planet's surface and approximately 96 percent of this water is contained in the major oceans of the world. The oceans drive our planet's weather, regulate its climate and provide us with breathable air, which ultimately supports every living creature on Earth.

The oceans are also vital to our global economy. They produce the food that billions of people depend on for survival, while also being a source of resources, including valuable medicines that treat a wide range of ailments and diseases. The oceans also drive local and regional economies through tourism. Every year, millions of travelers are drawn to coastal regions around the world to enjoy activities above the water and to explore what lies below the surface. Considering how important the oceans are to our way of life, it is incredible how little we know about what lies beneath their surfaces.

Coral reefs

The oceans include a wide range of different ecosystems, but perhaps the most frequently visited marine ecosystems of all are coral reefs. Coral reefs are known as the rainforests of the sea for good reason – they are one of the most diverse ecosystems on the planet, supporting about a quarter of all known ocean species. This figure is even more astounding when you consider that coral reefs comprise just a fraction of one percent of the ocean floor. They are also particularly vulnerable to degradation, given they are only found in a narrow window of temperature, salinity and depth.

Humans have studied the biology and physiology of corals for decades, but the underwater environment remains largely foreign to many people. Fact is, we have more accurate maps of the surface of Mars than we do of the seafloor. And guides of the marine environment suitable for recreational users are almost non-existent.

Reef Smart aims to change this situation. Our detailed guides seek to educate snorkelers and divers alike. Our goal is to improve safety and enhance the marine experience by allowing users to discover the unique features and species that can be found at each site.

Preserve and protect

Hopefully our guidebooks and handheld waterproof cards will help you get to know the underwater environment in general, and reefs in particular. We believe that the more people can come to appreciate the beauty of the underwater world, the more they will be willing to take steps to protect and preserve it.

The world's oceans are experiencing incredible pressures from all sides. Rising temperatures, increasing acidification and an astonishing volume of plastics that end up both in the water and in marine organisms are endangering these precious resources.

There are some big problems to overcome. But a better, more sustainable future is possible. Each and every one of us can make a difference in the choices we make and the actions we take. Together we can help make sure the coral reefs of this world are still around for future generations of snorkelers and divers to enjoy.

Sincerely, the Reef Smart team

A view of the Intracoastal Waterway from the air.

About Palm Beach County

Location and formation

Palm Beach County is in the southeastern part of the state of Florida, which itself is in the southeastern part of the continental United States. The county covers a total area of 1,977 square miles (5,120 square kilometers) not counting the area covered by Lake Okeechobee. Much of the central and western portions of the county are lightly developed and are earmarked for agriculture. The county is roughly 47 miles (75.5 kilometers) in length, stretching from the City of Jupiter in the north to Boca Raton in the south, and 52.5 miles (84 kilometers) east to west.

The state of Florida sits on the Florida Plateau, a geological formation dating back 530 million years. The plateau formed through a mix of volcanic activity and marine sedimentation. Many of Florida's features and soils, such as the central ridge that runs the length of the state and the two sand ridges near the coast of Palm Beach County, formed as a result of the alternating forces of deposition and erosion during a period when sea levels were much higher than they are now. Palm Beach County is relatively flat, with an average elevation of just 15 feet (4.5 meters) and a high point of 53 feet (16 meters) on the coastal ridge near Juno, located in the north of the county. The county's soil belies its marine origins, consisting primarily of sandy soils overlying porous karst limestone to the east, and organic "muck" associated with the Everglades and Lake Okeechobee to the southwest.

The history of Palm Beach County

Modern day Palm Beach County supports a large population and thriving economy along the Atlantic coast. But its history extends back tens of thousands of years to the early hunter gatherers that roamed North America.

Thomas Barrat/Shutterstock©

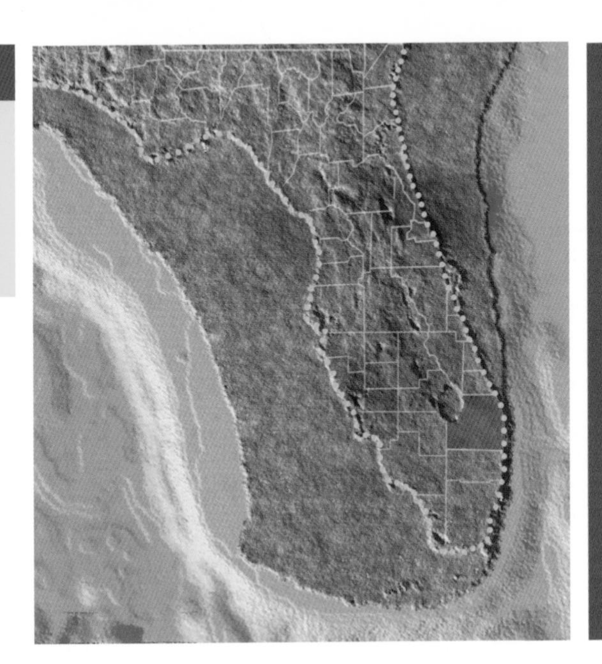

DID YOU KNOW? 

Tens of thousands of years ago, water was trapped in massive glaciers that stretched down from the poles. As a result, the ocean's surface was much lower than it is today. Almost the entire Florida Plateau was exposed to the air and covered in arid, scrub-like vegetation.

The current coastline of Florida is shown in yellow in the adjacent diagram, while Palm Beach County is in red. Note that the current position of Palm Beach's coastline is similar to where it was thousands of years ago.

Early history

Archaeologists have identified 10,000-year-old remains of hunter gatherers, called Paleoindians, in nearby Indian River County and in Jefferson County located in the north of the state. The oldest-dated archaeological site in Palm Beach County stretches back as far as 1000 BC – a set of burial and habitation mounds named Belle Glade, located in the Okeechobee Basin.

Anthropologists believe the early inhabitants of the area now known as Palm Beach County originally consisted of small tribes of hunter-gatherers who subsisted on a combination of fish, shellfish, deer, raccoon and plants such as sea-grape and prickly pear. The strongest of the tribes in this area were the semi-permanent Tequesta who lived in villages along the coast of what is now southern Palm Beach County down to northern Miami-Dade County. They were the second-most powerful tribe in southern Florida behind the Calusa, who controlled much of the southwestern part of the state. The Jeaga and the Jobe controlled the areas to the north of the Tequesta, stretching all the way up to the Indian River, while the Mayaimi inhabited what is now southwestern Palm Beach County.

The arrival of the Spanish explorer Juan Ponce de León in 1513 marked the beginning of the decline of these tribes, all of whom eventually perished from a combination of European diseases brought over by the Spanish and deaths from intertribal warfare, including fights with displaced tribes from up north who were forced south, particularly the Seminole Indians.

The United States acquired the area now called Florida from Spain in 1821. The state's first census took place in 1860, and it recorded a population of 34,730, with just 517 in southern Florida.

Recent history

Incorporated in 1909, Palm Beach County took a while to develop. The U.S. Army built Fort Jupiter in 1838 on the Loxahatchee River during the Second Seminole War – considered the first non-native permanent building in the county. The army also built a trail down to Fort Lauderdale, which would later become known as the Military Trail. In 1841, the first recorded name of a county resident was Seminole Chief Chachi, who oversaw a village located near where Palm Beach Lakes Boulevard and I-95 currently intersect.

The Jupiter Inlet Lighthouse entered service in 1860 with two keepers in residence. The first settlers around Lake Worth arrived in 1873, and the first hotels and schoolhouses were established within the next decade. Postmen established the Barefoot Mailman route down to Miami around this time, but the route was soon replaced by the railroad, which extended south from Jupiter, first to Juno and then eventually all the way down to Fort Lauderdale. Efforts to drain the Everglades, which began in the early 1900s, helped expand access to build-able land along the coast and spurred even faster growth.

Palm Beach relinquished control of some of its southern areas in 1915 to the newly formed Broward County and a part of its northern territory in 1925 when Martin County incorporated. In 1928, a hurricane destroyed much of the region, rupturing a dike around Lake Okeechobee and leading to the death of around 3,000 residents. The Great Depression struck on the heels of that natural disaster, but development in the area did not stall for long.

After the Second World War, many servicemen returned from the overseas conflict and chose to settle in the mild, sunny Florida climate after having spent their training days at the many military airfields built during the wartime effort. The county's population grew from 114,688 in 1950 to more than 836,000 in 1990 and over a million residents just a decade later.

Palm Beach County today

Population

Palm Beach County has a population of nearly 1.5 million residents. It is the third largest county in Florida by population – second largest by total area. The county's largest metropolitan area is West Palm Beach, which is located midway along the county's coastline. The second largest city is Boca Raton to the south and the third largest is Boynton Beach, which sits between West Palm Beach and Boca Raton. Palm Beach County also enjoys an influx of partial-year residents during the winter months, who are often referred to as "snowbirds" (See Did You Know? on page 12). Palm Beach County's economy is diverse and the main industries are tourism, construction and agriculture.

The basics

English is the official language in Florida, although Spanish is often heard in the southern parts of the state, including in Palm Beach County. The currency is the U.S. dollar, and there are plenty of bank branches, ATMs and foreign exchange outlets in the major cities and among the smaller towns. Development is most dense along the coast, with most of the southwestern portion of the county developed for agricultural use, primarily sugarcane fields which cover approximately one third of the county's total area.

Palm trees line many of the streets throughout Palm Beach County. Sean Pavone/Shutterstock ©

A view of downtown Palm Beach, set between the Atlantic Ocean and the Intracoastal Waterway.

The electricity in Palm Beach County is the North American standard 110 volts / 60 hertz with flat-bladed plugs and a rounded grounding pin. Internet is available at most hotels and in many coffee shops, eateries and other local businesses – sometimes free, sometimes paid.

Visitors

More than 8 million people visit Palm Beach County each year. They come to play golf, to kayak, horseback ride, surf, kite surf, fish and to relax on the county's 47 miles (75.5 kilometers) of sandy beaches. They also come for the fine

DID YOU KNOW

Every winter, Florida welcomes millions of "snowbirds," attracted by the sun, sand and mild winters of the "sunshine state." These individuals are usually retirees who travel from across the northern states of the U.S. and Canada. Incredibly, over 3 million Canadian snowbirds travel south each winter, equivalent to nearly one tenth of the country's population. The term "sunbird" has also started to catch on in recent years and refers to a reverse migration of southerners heading north.

Pisaphotography/Shutterstock ©

dining, the nightlife and the culture. The region boasts the Kravis Center for the Performing Arts, the Dolly Hand Cultural Arts Center and the Palm Beach Opera, along with dozens of museums. The proximity of the Gulf Stream Current ensures warm waters year-round. With a multitude of wrecks and long stretches of reef and ledge to explore, Palm Beach County is also a very popular location for diving and snorkeling, among visitors and locals alike.

Getting there and getting around

Getting there

Palm Beach County's main airport is the Palm Beach International Airport (PBI) located midway along the coast, just a few miles west of downtown West Palm Beach. Nearly 7 million passengers pass through the airport each year. Visitors have the convenience of also choosing to fly into the Fort Lauderdale-Hollywood International airport (FLL), located 30 miles (48 kilometers) to the south. Miami International Airport (MIA) is also an option when paired with a trip on the regional Tri-Rail train.

Getting around

Palm Beach County has an extensive community public transit system that covers most of the developed eastern areas of the county as well as the western urban area. Palm Tran maps and information are available on the county's website: **Discover.pbcgov.org/palmtran**. Bus fares are affordable at just $2.00 for a one-way trip, but multi-day passes are also available. The transit system also links up with the equivalent system in nearby Broward County, known as BCT. All major car rental agencies operate in the county, and finding parking is not too challenging. Traffic can be difficult during the peak rush hours of the morning and late afternoon, however, so plan your trips accordingly. Ride-sharing programs such as Uber and Lyft are also active in the county.

Environment

Weather

Palm Beach County enjoys mild winters and relatively hot summers. The average summer high is around 89°F (32°C), while temperatures in winter are a comfortable 74°F (23.5°C). The county receives an average of 62 inches (157 centimeters) of precipitation per year, with the bulk of that happening during the summer months. Storms do not typically last long, and rain clouds move through the region quickly. Florida is subject to occasional hurricanes, however, and the hurricane season traditionally stretches from June 1 through November 30. The county has been hit by dozens of hurricanes over the years, with one of the most harmful being the unnamed storm in 1928 that caused 3,000 deaths.

The sea temperature along the coast varies from an average low of 71.5°F (22°C) in February up to a high of 87°F (30.8°C) in August. The Florida Current (the start of the famous Gulf Stream Current) runs very close to shore along the county's coast. It can get to within as little as a mile of the shore, which can make diving a little more challenging on some of the deeper, offshore wrecks. Other times, the strong northerly current is nearly 10 miles (16 kilometers) off shore.

Waves and visibility

The coast of Palm Beach County is exposed to the Atlantic Ocean; there are almost always waves crashing on the beach. If a storm is off shore, or the winds are blowing from the east, the seas can easily top 8 feet (2.5 meters) which makes boat diving a no-go but it can make for great surfing and kite surfing conditions. High winds are most common in February, but they occur only periodically. The ocean is generally calm enough to allow boat dives, and even shore dives for those intrepid enough to wade through the breakers.

Snorkeling at this time can be a challenge, however, as the rough surf on the beach can kick up a lot of sand making for poor visibility in shallow waters. There is plenty of sand along the coast for a storm to stir up in the water. But in general, visibility in deeper water ranges from 50 to 60 feet (15 to 18 meters), depending on the site and the time of year. It is not uncommon for visibility to reach as much as 100 feet (30 meters), but neither is it unheard of to have visibility as poor as just 5 feet (1.5 meters). When the Gulf Stream Current comes in closer to the shore, the visibility improves, but the currents are stronger.

Visibility also declines substantially in the waters surrounding the three main inlets during an outgoing tide. Dark water from the Intracoastal Waterway flows through

Andrea Whitaker ©

Divers from Pura Vida signal their boat for a pickup.

Zachary R. Nolan ©

Lobsters are frequently spotted along the county's many ledges and patch reefs.

the inlets and is carried northward with the prevailing currents. Most operators plan their diving schedules to accommodate the tidal cycle for this reason.

Currents and tides

The tidal cycle along this stretch of coast is modest compared to many other parts of the world, with the difference between high and low tide measuring less than 3 feet (1 meter) along the coast and in the Intracoastal Waterway. The Florida Current is the major driver of ocean circulation in these parts. It is a well-defined component of the larger, and better-known Gulf Stream Current. The Florida Current enters the Gulf of Mexico before returning along the coast of Florida, passing through the Straits of Florida and rejoining the rest of the Gulf Stream to head in a northeastern direction toward Europe. The Current drives most of the system off the coast of Palm Beach County, leading to inshore currents that tend to run parallel to the shoreline, either northward or southward, depending on how close the Gulf Stream is to shore, among other factors.

Ecosystems

Coral reefs

Southeast Florida has a set of fringing reefs that stretch from Palm Beach County in the north down to Miami-Dade County in the south. Known collectively as the Florida Reef Tract, these reefs include three distinct, parallel reef tracts – an inner, middle and outer reef, sometimes referred to as first, second and third, respectively. Florida is the only state in the continental U.S. where divers and snorkelers can explore a coral reef from shore.

All three reefs are distinctly visible in Broward County to the south, but only the outer reef continues north much past Boca Raton in Palm Beach County. It terminates just off shore from West Palm Beach at the site known as Turtle

DID YOU KNOW? 

Several species of sea turtle are found in Florida waters, including loggerhead, green, hawksbill, Kemp's ridley and leatherback sea turtles. From March until the end of October, it is estimated that there are over 70,000 sea turtle nests, most of them belonging to loggerheads – the most common species.

On Mondays, Tuesdays, Wednesdays and Fridays (June through July 14) the John D. MacArthur Beach State Park organizes tours to watch sea turtles nest on the park's beach. Each season this stretch of beach sees over 2,000 sea turtle nests. Register online at **Macarthurbeach.org**. Tickets go on sale starting May 30 for $12.00 each.

Mound. To the north, the seafloor spreads out into a wide area of ledges and sections of reef that no longer form a distinct reef tract.

The reefs are home to a diverse assemblage of stony corals, sponges and reef fish species. They are "relic" reefs, meaning they are no longer actively growing upward. They date back to prehistoric times, roughly around the end of the Pleistocene era, which lasted from 2.6 million years ago to just 11,700 years ago. At that time, the world's sea levels were much lower than they are today due to the massive amount of water trapped inside that era's many glaciers. Today, the reefs have been recolonized with living corals and other marine organisms, which makes for great diving and snorkeling, particularly drift diving as the prevailing currents typically run parallel to the reefs and ledges found in the area.

Inner reef

The inner reef is difficult to find along the Palm Beach coastline. The Florida coast drifts eastward as it heads north into Palm Beach County, largely engulfing this component of the reef tract. However, there are many small fringing reefs found adjacent to shore along the county's coastline – these reefs are not part of the reef tract, but they do make for great snorkeling.

Loggerhead Marinelife Center ©

A turtle release event organized by the Loggerhead Marinelife Center.

Middle reef

The middle reef sits just inside the outer reef tract, cresting at a depth of 45 to 50 feet (13.5 to 15 meters). Researchers believe it stopped growing upward nearly 3,700 years ago, and core samples suggest it consisted primarily of massive corals. It is associated with a ridge that may have once been a shoreline as sea levels rose. This component of the reef tract disappears off shore of Boca Raton, although patches of it persist a little farther to the north.

Outer reef

The outer reef is the deepest of the three reef tracts, cresting at a depth of 50 to 55 feet (15 to 17 meters) below sea level, and the target of much of the reef diving in the county. The reef stopped growing 8,000 years ago – shortly after sea levels started rising due to melting glaciers. Core samples indicate it consisted of acroporid corals, involving a mix of elkhorn and staghorn corals. This component of the reef tract maintains the familiar spur and groove formations of a fringing reef on its eastern edge, suggesting it was once a well-established, relatively shallow shoreline reef during the Pleistocene era. Gaps in the outer reef suggest the presence of river outlets that may once have eroded a channel through the substrate, physically limiting the establishment of the corals.

Artificial reefs

Palm Beach County's artificial reef program got its start in the 1960s, but lagged other regional reef programs until 1985, the year after neighboring Broward County purchased the Mercedes I to sink as an artificial reef along its coastline. The vessel had run aground on the Palm Beach county coastline and become a media sensation. But without an artificial reef program in place, Palm Beach could do little to leverage the story for its own benefit.

The county's artificial reef program has deployed 55 vessels, 100,000 tons of concrete, and 133,000 tons of limestone boulders. The program is part of the county's Department of Environmental Resources Management. Many of these artificial reefs are highlighted in the pages of this book. The reef program also works with volunteers to survey and monitor both natural and artificial reefs in the region. This effort is aimed at collecting data to inform management decisions. For more information, visit: **Pbcreefteam.com**

Marine management, research and conservation

Coral reefs are valuable resources for the regions in which they are found. In Palm Beach County, coral reefs provide around 4,500 jobs through activities such as fishing, diving and snorkeling, and generate income of $141 million annually. As such, many Floridians are passionate about the environment and take the management and protection of their coral reef ecosystem very seriously. This passion is on display in the many conservation-oriented organizations that operate in Palm Beach County.

Loggerhead Marinelife Center

This rescue and rehabilitation center is located in Juno Beach, right off U.S. Hwy 1. The center's stated mission is to "promote conservation of ocean ecosystems with a special focus on threatened and endangered sea turtles." The 12,000-square-foot (1,115-square-meter) eco-friendly facility houses an exhibit hall, outdoor classrooms, research labs and a full-service veterinary hospital. The campus sits right on the ocean, and features a beach with lifeguards on duty, nature trails and picnic pavilions. Depending on the time of year, the center offers visitors a variety of experiences, including guided turtle beach walks in the summer and tours of local ecosystems throughout the year. The center is open daily from 10am to 5pm except major holidays. For more information and to plan your visit, check out their website: **Marinelife.org**

Gumbo Limbo Nature Center

The Gumbo Limbo Nature Center is located on Ocean Boulevard, due west of Boca Raton, on the protected western edge of the barrier island. The center is a joint initiative by the city of Boca Raton, Greater Boca Raton Beach and Park District, Florida Atlantic University and the Friends of the Gumbo Limbo. The 20-acre site offers visitors the chance to walk along a raised boardwalk in the middle of a natural forest (open seven days a week from 7am to dusk) as well as access the center's education, outreach and wildlife rehabilitation programs from Monday to Saturday, 9am to 4pm. There are aquariums, rehabilitating sea turtles and even many rare and endangered species that seek shelter in the center's natural settings. Visitors can learn more about the center or find out what programs are happening while they are in the area by checking online: **Gumbolimbo.org**

ECO TIP

Many of the organizations listed in these pages are constantly on the lookout for volunteers to support their missions and to help achieve their goals. Age, skill and experience are not always important – in many cases all you need is a passion for the marine ecosystem and a desire to help.

If you are interested in helping to make a difference in the effort to preserve and protect our oceans and marine-related ecosystems, please reach out to any of these organizations that are doing work that interests you. You can learn how to be a better advocate through volunteering. When added together, even small steps can make a big difference.

Palm Beach Zoo

The Palm Beach Zoo and Conservation Society located at Dreher Park in West Palm Beach has been around for more than 50 years. The zoo houses more than 500 animals on 23 acres of land. It offers many on-site education and conservation programs and is open daily from 9am to 5pm except American Thanksgiving Day and Christmas Day. For more information, visit: **Palmbeachzoo.org**

Andrea Whitaker ©

Goliath grouper are common off Palm Beach's coast, where they congregate during summer spawning events.

South Florida Science Center and Aquarium

Those visitors looking to experience aquatic life without getting wet can visit the science center and its 10,000-gallon (37,850-litre) installation, Aquariums of the Atlantic, located in Dreher Park in West Palm Beach. The aquarium features native species as well as many of the invasive species that are wreaking havoc on local habitats, including the invasive lionfish. From touch tanks to coral reefs and even the Everglades ecosystem, there is plenty for visitors to experience at the science center and aquarium. For more information, visit: **Sfsciencecenter.org**

Florida Manatees

One of Florida's most fascinating underwater creatures is the Florida manatee (*Trichechus manatus latirostris*), which is a subspecies of the West Indian manatee, and a marine mammal. Manatees are also, according to many, the origin of the myth of the mermaids – hence the name of their order, Sirenia (after the sirens of Greek mythology).

Manatees are protected from capture, harm and harassment by the Marine Mammal Protection Act of 1972. The slow-moving herbivores are often seen near the surface of the water, which helps explain why boat collisions are the main cause of death, killing over 80 individuals each year. Manatees frequent the many inlets and bays of the region, and in cold weather, they often congregate in the warmer waters of rivers and springs – including the discharge outlets from area power plants. Visitors are most likely to see manatees in Palm Beach County during the winter months (November through March), with the potential of seeing hundreds at a time at the Manatee Lagoon, located just off Hwy 1 in West Palm Beach, adjacent to the Florida Power & Light power plant. A nearby educational center provides visitors with more information about the biology and conservation efforts of these gentle giants nicknamed "sea cows." For more on the Manatee Lagoon, visit: **Visitmanateelagoon.com**

John D MacArthur Beach State Park

This is the only state park in Palm Beach County. Established in 1989, the park consists of 438 acres of pristine Florida habitat located in North Palm Beach, just north of Singer Island. It stretches from the Lake Worth Lagoon across the barrier island to the open ocean. The park includes 1.6 miles (2.5 kilometers) of sandy beach that offers visitors the chance to swim, snorkel and dive in the ocean. There is no lifeguard at this beach, and dive flags are required here.

The park shelters 22 endangered or threatened species, which visitors can potentially spot as they walk along the many nature trails or kayak through the sheltered waterways. They can also learn about the biological riches that the park was established to protect, through the displays in the Nature Center. There, visitors will also learn about the resident Jaega Indians who lived here many centuries ago, as indicated by the kitchen middens they left behind. The park is open from 8am to sundown and is accessible with a $5 per vehicle fee. It is a place to unwind and find nature in the middle of an otherwise developed stretch of coastline. For more information on the park visit: **Macarthurbeach.org**

Palm Beach County beaches

Beaches run along most of the length of Palm Beach County's 47 miles (75.5 kilometers) of coastline. Officially, there are 12 municipal beaches and 14 county-run beaches with lifeguards on duty. That said, all beaches in Florida are public, and you have the right to walk up and down the beach provided you do not venture above the mean high-tide mark. Beach "access" is a more complicated topic, however. The beach is accessible via the many parks that dot the coastline, along with dozens of specific public right-of-way access points. Parking at state, county and municipal parks is adequate, although it gets busy when the beaches get busy during holidays and high season.

With so many beaches to choose from, there is bound to be a spot for everyone regardless of how busy things get. To check on the beach conditions before you head out, visit: **Thepalmbeaches.com/beach-conditions**

From north to south, the guarded beaches include:

1. Coral Cove Park
2. Dubois Park
3. Jupiter Beach Park
4. Carlin Park
5. Ocean Cay Park
6. Juno Beach Park
7. Loggerhead Park (associated with the Loggerhead Marinelife Center)
8. Ocean Reef Park (described on page 106)
9. Phil Foster Park (described on page 126)
10. Riviera Beach/Singer Island Municipal Beach
11. Palm Beach Shores Beach
12. Peanut Island
13. Palm Beach Midtown Municipal Beach
14. Phipp's Ocean Park Municipal Beach
15. R.G. Kreusler Park
16. Lake Worth Municipal Beach
17. Lantana Municipal Beach
18. Ocean Inlet Park (described on page 184)
19. Boynton Beach Municipal Beach
20. Gulfstream Park
21. Delray Municipal Beach
22. Atlantic Dunes Municipal Beach
23. Spanish River Park Municipal Beach
24. Red Reef Park Municipal Beach (described on page 220)
25. South Beach Park Municipal Beach
26. South Inlet Park (described on page 224)

Relaxing on the county's many beaches is a popular pastime. Peera_stockfoto/Shutterstock ©

SAFETY TIP

Even many Floridians are confused about beach access rights. Legal cases involving access and public rights-of-way are plentiful, with some disputes stretching out over many years. Many residents have installed gates and signage indicating that trespassers will be prosecuted. It is possible some of these access points may in fact be public, while others that appear public may in fact be for the exclusive use of local residents. To avoid spoiling a vacation with a lengthy discussion about policy, visitors may be better off accessing the sand and surf via the many public beaches. These beaches also typically offer shower and bathroom facilities, which is a definite plus when spending a day swimming and relaxing on the sand.

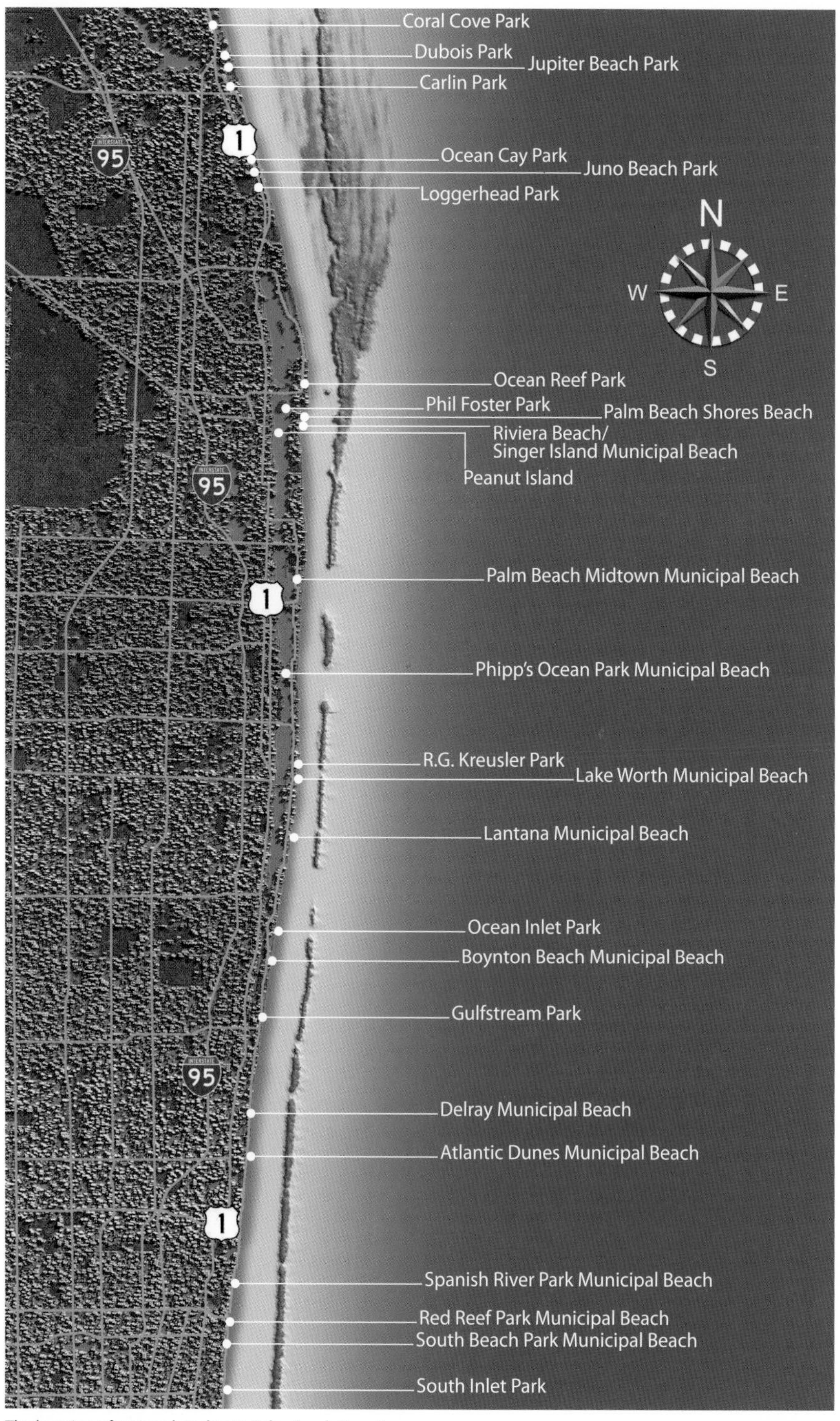

The location of various beaches in Palm Beach County.

In case of emergency

There are many organisms that can put a damper on your experience in Palm Beach County, from jellyfish to stingrays to the invasive lionfish that have installed themselves in the reefs and wrecks throughout Florida's waters. Many of these potentially dangerous marine creatures are listed in a special section toward the back of the book (pages 258–261). We have included information in that section on the harm these species can cause and some of the common treatments that might help. NOTE: This book is not intended as a substitute for professional medical help.

If you are unlucky enough to end up injured after diving, snorkeling, surfing or otherwise enjoying the reefs of Palm Beach County, there are many clinics and hospitals you can visit to receive top-notch care. Because no matter how careful you are, accidents can happen.

If your injury is not an emergency, consider visiting the nearest walk-in clinic or urgent care clinic during its normal operating hours – they will vary by clinic, but many are open between 8am and 8pm. If it is an emergency, however, call 911 or go to one of the regional emergency rooms as quickly as possible.

The healthcare system in the U.S. is very efficient, but it can also be expensive. If you are visiting from overseas, it is advisable to have some form of health insurance coverage prior to your arrival in the U.S. Without coverage, an accident could potentially be very costly.

Emergency contacts

Police 911
Hospital and Ambulance: 911
Fire department: 911
Coast Guard: 954-927-1611
(Radio channel 16)

Hospitals/Emergency rooms

Cleveland Clinic – Florida (8am-5pm weekdays)
4520 Donald Ross Blvd, Suite 200, Palm Beach Gardens
561-904-7200

Palm Beach Gardens Medical Center (E.R.)
3360 Burns Rd, Palm Beach Gardens
561-622-1411

Divers practice proper safety during a training session.

JFK E.R. – Palm Beach Gardens (24hr E.R.)
4797 PGA Blvd, Palm Beach Gardens
561-548-8200

JFK Medical Center – North Campus (24hr E.R.)
2201 45th St, West Palm Beach
561-863-3900

St. Mary's Medical Center (24hr E.R.)
901 45th Street, West Palm Beach
561-844-6300

Good Samaritan Medical Center (24hr E.R.)
309 N Flagler Dr, West Palm Beach, FL 33401
561-655-5511

JFK Medical Center – Atlantis (24hr E.R.)
5301 S Congress Ave, Atlantis
561-965-7300

Bethesda Hospital East (24hr E.R.)
2815 S Seacrest Blvd, Boynton Beach
561-737-7733

JFK Emergency Room – Boynton Beach (24hr E.R.)
10921 S Jog Rd, Boynton Beach
561-548-8250

Bethesda Hospital West (24hr E.R.)
9655 W Boynton Beach Blvd, Boynton Beach
561-336-7000

Delray Medical Center (24hr E.R.)
5352 Linton Boulevard, Delray Beach
561-498-4440

Boca Raton Regional Hospital (Urgent Care)
10 E Palmetto Park Road, Boca Raton
561-955-2700

Decompression / Hyperbaric chambers

St. Mary's Medical Center (open 24hr)
901 45th Street, West Palm Beach
561-882-2852

Surfing

Surfing in Palm Beach County can be truly exceptional – one might argue it has the best surfing spots along the whole of the state's southern shoreline. Conditions need to be just right for the best breaks, however, and there are some accessibility challenges.

The Bahamas shields much of Southeast Florida from westerly swells, but Palm Beach County is located adjacent to the northern end of the island chain. Therefore, swells tend to wrap around the islands and hit the county's top beaches. The county is also the easternmost part of Florida, which makes it well-positioned to receive swells from the north.

The result is that Palm Beach County boasts many surf spots that, on a good day, rival some of the more recognized and renowned breaks elsewhere in the U.S., especially during late fall and winter. Of particular note is Reef Road, which is widely regarded as one of the best surfing spots in the entire state.

Our three-star rating system indicates the quality of the site in most conditions. And while the wave varies from day to day, we have attempted to provide as much information as we can about its quality, consistency and the direction of the break, to help you get the most out of your surfing experience.

We have also provided information on access to the break and local conditions, including rip currents and undertows. Be sure to check up-to-date conditions online at any one of the many surf report sites available, such as: **Magicseaweed.com** and **Surfline.com**.

Surfing at dawn as the sun rises on the east coast.

sw_photo/Shutterstock©

SAFETY TIP

Rip currents flow from the shore to the open sea and can be incredibly strong, particularly in stormy weather and at low tide. Surfers use rip currents as a quick and easy passage through the waves to the lineup. But for swimmers – particularly weak swimmers – they can be extremely dangerous and are known to kill approximately 100 people every year in the United States. Rip currents can be difficult to spot, but there are several tips that can improve safety:

Check for warning flags before entering the water – red flags are placed on the beach when conditions are dangerous.

Ask a lifeguard, if present, about known rip currents in the area.

Scan the water for signs of a rip current before entering. Areas without breaking waves, or with cloudy or choppy water, often indicate rip currents.

Observe any floating objects, such as bits of wood or seaweed. Areas where floating objects are washed out to sea are often rip currents.

If you are caught in a rip current, always swim perpendicular to the current to free yourself, rather than against it.

Surfing and other services

Blueline Surf & Paddle
997 Florida A1A, Jupiter
Tel: 561-744-7474
Email: info@bluelinesurf.com
Bluelinesurf.com

Ocean Magic Surf & Sport
103 US-1 C-6, Jupiter
Tel: 561-744-8925
Email: oceanmagicsurf@gmail.com
Oceanmagicsurf.com

Ground Swell Surf Shop
11 Donald Ross Rd, Juno Beach
Tel: 561-622-7878
Email: info@groundswellsurfshop.com
Groundswellsurfshop.com

Get Wet Surf Shop and Skate Shop
237 Blue Heron Blvd #2, Riviera Beach
Tel: 561-253-8754
Getwetsurfandskate.com

Crowd Control Surf Co.
6687 42nd Terrace N unit B, Riviera Beach
Tel: 561-723-8039
Crowdcontrolsurfstore.com

P.B. Boys Club
307 S County Road, Palm Beach
Tel: 561-832-9335
Facebook.com/PB-Boys-Club-21362396529

Gypsy Life Surf Shop
1401 Clare Ave, Suite 300, West Palm Beach
Tel: 561-313-5675
Email: hello@gypsylifesurfshop.com
Gypsylifesurfshop.com

Island Water Sports
513 Lake Ave, Lake Worth
Tel: 561-588-1728
Email: iwslw@aol.com

Nomad Surf Shop
4655 N Ocean Blvd, Boynton Beach
Tel: 561-272-2882
Nomadsurf1968.com

The Drop In Surf Shop
220 NE 1st Street #900, Delray Beach
Tel: 561-562-5545
Email: info@surf-district.com
Surf-district.com

Epic Surf Shop
1122 E Atlantic Ave # E, Delray Beach
Tel: 561-272-2052
Email: epicwalt@hotmail.com
Epicsurfshop.com

Waves Surf Academy
4800 N Federal HWY Suite# B100, Boca Raton
Tel: 561-843-0481
Email: ericdernick@wavessurfacademy.com
Wavessurfacademy.com

Boca Surf and Sail
3191 N Federal Hwy, Boca Raton
Tel: 561-394-8818
Email: bocasurfandsail@yahoo.com
Bocasurfandsail.com

Surf breaks

Juno Pier to Jupiter Inlet

Waves break on both the north and south sides of Juno pier, depending on the direction of the swell and the position of the sandbars in the water. This spot is not known for its big waves, but under the right conditions it can generate overheads and even double-overheads – usually during hurricane swell in the fall. Juno Pier is also one of the most consistent waves in Jupiter, producing decent left- and right-handers irrespective of the tide. This spot can get crowded as a result, particularly on weekends, which can lead to a certain amount of "localism." To the north of Juno Pier is a stretch of beach that runs uninterrupted for 3.5 miles (5.6 kilometers) to Jupiter Inlet. This stretch is dotted with plenty of great beach breaks, including Corners, Civic Center and Jupiter Inlet itself.

DID YOU KNOW?

When surfing, it is worth remembering drop-in etiquette to avoid being on the receiving end of some "choice language" from other surfers.

Remember that the surfer closest to the curl – the breaking edge of the wave – has priority over those farther out and the surfer already riding the wave has priority over paddling surfers. Dropping-in on a surfer from the outside and snaking a wave by cutting in from the inside are both considered bad form and can be dangerous.

While it can be tempting to try to catch what seems like the perfect wave, it is important to wait your turn in the line-up – there are enough waves for everyone.

Name:	Juno Pier	**Level:**	Intermediate
Location:	Jupiter	**Best tide:**	All
Wave direction:	Left and right	**Best season:**	Fall and winter
Wave length:	Medium	**Popularity:**	Medium to high
Sea bed:	Sand		
Type:	Consistent and mellow beach break		

Reef Road

Without a doubt, Reef Road is one of Florida's best surf spots. Located just south of the Lake Worth Inlet (Palm Beach Inlet), this site produces long left-handers that attract surfers from far and wide. This is undoubtedly a big-wave spot, which can produce long and consistent barrels that seem to go on forever. Reef Road requires a 6-foot (2-meter), north-to-northeast swell to really start firing but can grow to a terrifying 20-foot (6-meter) wave under the right conditions – usually during hurricane season or in the winter months. The downside of surfing Reef Road is the limited parking and tricky access. Ultra-expensive properties largely block access to the beach adjacent to the break; surfers must therefore park a mile or so south of the site where there is limited free parking around the Palm Beach Country Club, or take an even longer hike up the beach from the parking meters farther to the south. Trespassing through private beach gates, parking illegally, or even stopping for too long around Reef Road could land you a steep fine, so be warned.

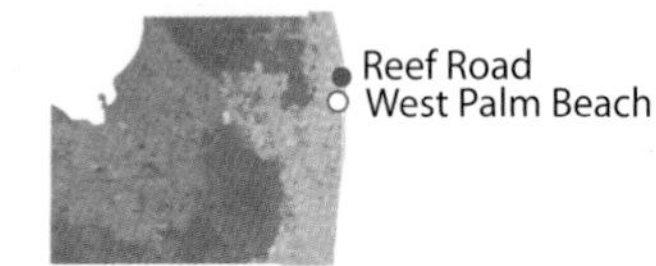

DID YOU KNOW? 

There is a sand-pumping house on the north side of the Lake Worth Inlet (also called the Palm Beach Inlet). This is the easternmost point in the state of Florida and is the aptly named Pump House surf spot.

This killer break only fires a handful of times each year, but it produces epic rides when the conditions are right, including huge, fast-breaking barrels that often require a jet ski-assisted tow-in.

Pump House is a dangerous place to surf, however. It is usually only attractive to professionals and those fearless enough to give it a try. Even so, it holds a legendary status for those surfers with the stomach for big waves.

Name:	Reef Road	**Level:**	Intermediate to advanced
Location:	Palm Beach	**Best tide:**	All, but particularly low
Wave direction:	Left	**Best season:**	Fall and winter
Wave length:	Long	**Popularity:**	Medium
Sea bed:	Sand and reef		
Type:	Big, long, powerful barrels		

Lake Worth Pier

Lake Worth Pier is one of the most consistent surf breaks in the county, which also makes it one of the most popular surf spots. Decent, spilling breakers become long, powerful barrels as the swell size builds. Waves can break on either side of the pier, but the south side produces lefts and rights and is often better-shielded from the wind. However, the north side is arguably the better surfing experience when conditions are perfect, producing fast and hollow right-handers. Do not "shoot the pier" by taking a right-hander from the south side through the pier to the north. Not only is it dangerous, it is also a good way to irritate the lifeguards. The site has plenty of metered parking, freshwater and other facilities, which on top of the consistency of the waves, draws surfers from far and wide. Unfortunately, the popularity of this spot has led to conflict, and over the years Lake Worth Pier has developed the dubious reputation of being the most "localized" surf spot in the county. As such, it is not a good option for beginners. The bad rep comes largely from a number of violent incidents that occurred decades ago, but it is still considered a bit of a "locals'" spot, so keep that in mind and remember your surfing etiquette when visiting.

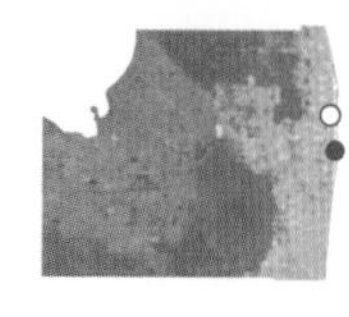

RELAX & RECHARGE

It is hard to find a more scenic spot for a drink and a bite to eat than **Benny's On The Beach**. This pier-based restaurant has been serving up quality food since 1986, but it is their fresh seafood that has been the foundation of their reputation.

Many people head to Benny's for breakfast, which is served from 7am, as the sun rises above the ocean. A number of their breakfast options are served with shrimp and lobster.

Benny's also has a varied lunch and dinner menu. The Crab Cake Sammy, which is lump crab cake topped with smashed avocado, crumbled bacon and red onion served on a fresh challah bun with lettuce, tomato and house horseradish sauce, is as good as it sounds. Visit: **Bennysonthebeach.com**

Name: Lake Worth Pier
Location: Lake Worth
Wave direction: Left and right
Wave length: Long
Sea bed: Sand and reef
Type: Spilling breakers and long powerful barrels
Level: Intermediate to advanced
Best tide: All
Best season: Fall and winter
Popularity: High

Boynton Inlet

Waves can break on either side of the Boynton Inlet. The south side is usually the best in north to northeast swells, whereas the north side can produce rideable waves in a southeast swell.

Waves here are often waist-high and above, and more experienced surfers will be able to get the most out of them. Watch out for the rocks at the waterline, the extent of which may be hidden from view from the shoreline.

For many surfers, this spot does not have the consistency of other nearby locations. However, the beach is easy to find and there is plenty of free parking.

There are also showers, picnic benches and even a playground for surfers who have kids, so if you want a place that is low key, less crowded and easy to access, Boynton Inlet might just be your ideal spot to surf in Palm Beach County.

SAFETY TIP

Several different species of shark are found in Florida waters, including tiger sharks and bull sharks. Although attacks on swimmers, surfers, snorkelers and even divers do occasionally occur, it is important to keep in mind that deliberate attacks by sharks are very rare. Most shark bites are accidental, brought about by the limited visibility that exists close to shore. On average there is fewer than one fatality each year from shark attacks in the United States.

Consider that statistic relative to the following: According to data provided by the Consumer Product Safety Commission, more than twice as many people die annually in the U.S. from vending machines falling on top of them. Moreover, 20 people are killed by cows each year, according to the Centers for Disease Control and Prevention.

Name:	Boynton Inlet	**Level:**	Intermediate to advanced
Location:	Boynton Inlet	**Best tide:**	Low
Wave direction:	Left	**Best season:**	Fall and winter
Wave length:	Medium	**Popularity:**	Medium
Sea bed:	Sand and reef		
Type:	Spilling breakers and barrels		

Delray Beach

Delray Beach boasts 2 miles (3.2 kilometers) of public beach, easily accessible from Hwy A1A, which runs directly along the edge of the ocean, for the most part. Surfable waves break all along the beach, but the most rideable waves are often near Anchor Park in the vicinity of the *SS Inchulva* shipwreck and even further south at Atlantic Dune Park. Both of these spots are beach breaks that deliver decent surf on every tide and swell combination.

There are five metered parking areas, as well as paid parallel parking spots along Ocean Blvd, but the place can get jammed when conditions are ideal. Delray Beach has a nice mellow vibe overall, which dates back to the early days of surfing in the 1950s. Today, surfers are often outnumbered by kiteboarders as the beach has become a popular spot for kiteboarding, with lots of room to launch kites and plenty of bars and restaurants to enjoy throughout the day and into the evening.

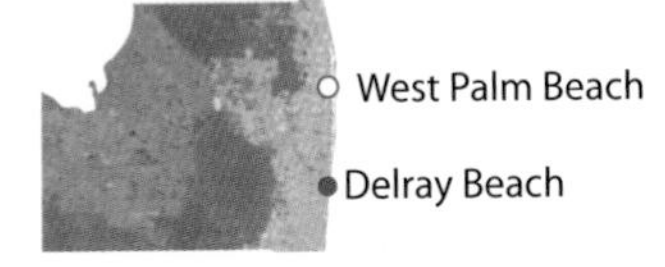

ECO TIP

Most people consider surfers to be an eco-friendly bunch but some of the products they use are far from good for the environment. For example, almost all the board wax available on the market today is made from petrochemical-based paraffin. Natural and organic waxes made from various plants and beeswax are now starting to appear, and surfers are taking notice.

Surfers also tend to pile on the sunscreen, which studies have shown often contains ingredients that can damage coral reefs. Biodegradable sunscreens that lack ingredients such as parabens, cinnamate, benzophenone and camphor derivatives have been around for a while now. It is always worth taking time to read the label to ensure your selection is good for the environment; it usually means it is better for your body too.

Name:	Delray Beach	**Level:**	Intermediate
Location:	Delray Beach	**Best tide:**	All
Wave direction:	Left and right	**Best season:**	All year, even summer
Wave length:	Medium	**Popularity:**	Medium to high
Sea bed:	Sand		
Type:	Spilling breakers and barrels		

Boca Raton South

Boca Raton may fall short of being paradise for surfers, but the stretch of coastline that runs from Jap Rock in the north through Spanish River and down to Red Reef Park in the south can produce some mighty fine waves under the right conditions. There are sandbars all along this 2-mile (3.3-kilometer) stretch of coast, but also multiple reef areas, which adds an element of risk not present at other sites in Palm Beach County.

Boca Raton is highly sensitive to swell size and direction – north to northeast swells are needed to generate the right kind of wave, which is unfortunately quite rare along this stretch of coast. But under the right conditions, Boca will produce a long, consistent head-high wave with a quality rarely found in Floridian waters. At Jap Rock, which is arguably the best end of Boca to surf, waves follow the contour of the reef, which sprouts out from the beach into the water like a giant mushroom.

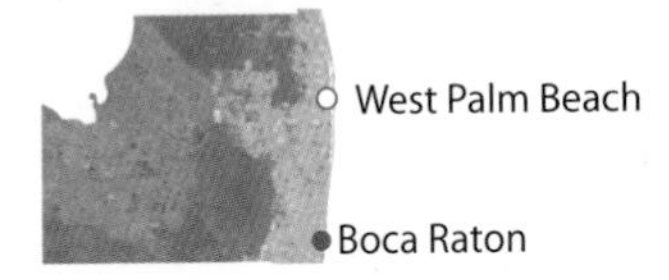

SAFETY TIP

If you are concerned about sharks, there are several things you can do to lower your chances of a negative encounter:

Check the surface for dorsal fins before entering the water and avoid areas where sharks are obviously present.
Do not enter the water if you already have an injury that causes blood to enter the water.
Avoid surfing at dawn, dusk and at night when shark feeding activity may be higher and visibility tends to be lower.
Avoid wearing high contrast colors such as orange and red, but particularly yellow, which has been shown to attract sharks.
Use free apps such as Dorsal, which allows users to document and share shark sightings with the online community.

Name:	Jap Rock, Spanish River, Red Reef	**Level:**	Intermediate
Location:	Boca Raton	**Best tide:**	All
Wave direction:	Left and right	**Best season:**	Fall and winter
Wave length:	Medium to long	**Popularity:**	Medium
Sea bed:	Sand and rock		
Type:	Spilling breakers and barrels		

Diving and snorkeling

Palm Beach County offers an incredible range of diving and snorkeling experiences. The region boasts a variety of natural and artificial reefs, with many of the latter strategically positioned around the area's three active inlets. Some of the county's artificial reefs are readily accessible from shore – including the world-renowned Blue Heron Bridge dive site and Phil Foster snorkel trail – while others lie a mile (1.6 kilometers) or more off the coast. The county's wrecks offer something for everyone, ranging in depth from less than 10 feet (3 meters) to more than 200 feet (61 meters).

The region's natural reefs offer incredible drift dives for divers of all experience levels. Those in the south are part of a semi-contiguous reef tract that extends northward from Broward County. In the north of the county, however, the reef tract breaks down into a broad swath of natural ledges and rocky reefs that cover a large portion of the offshore waters and create a varied mix of habitat for divers.

There are multiple snorkeling opportunities along the Palm Beach coastline, including sites associated with many county and municipal parks that offer a surprising level of biodiversity for sites that sit so close to shore.

We highlight 31 of the most popular dive and snorkel sites in this guidebook, including a few multi-wreck treks. We also provide a complete list of the established dive sites for the region so that visitors have a chance to grasp the true variety available. For each dive or snorkel site selected for this guidebook, we provide the information necessary to understand where the site is located, how long it takes to get there either by boat from the nearest inlet or by car from the nearest city center, and what divers and snorkelers can expect from their time in the water. For wreck dives, we include a detailed description of the history and story of the wreck, if known, to provide some interesting context for divers and snorkelers. Our three-

Zachary R. Nolan ©

Many of the artificial reefs in Palm Beach County are heavily colonized by corals.

SAFETY TIP

Since 2014, Florida law requires divers must "prominently display" a diver-down flag when diving. This alerts boat operators who encounter such a flag to take specific actions to avoid divers in the water.

The law allows some flexibility in how the dive flag is displayed. Vessel operators can display a flag from their vessel when divers are in the water, while divers can tow a traditional dive flag or a dive buoy to help mark their location on the surface. Divers in a group must be within 100 feet (30.5 meters) of the flag/buoy during their dive and vessels must remain 300 feet (91.5 meters) from divers in the water.

Most of the dive sites in Palm Beach are explored as drift dives, which makes the use of a towed buoy necessary since dive boats are not moored to the site. Many operators also require divers to have their own surface marker buoys to inflate upon surfacing for pickup by the boat.

Eric McDonald ©

Dive boats ensure that dive flags are very visible when divers are in the water.

star rating system offers insight about the difficulty level, strength of the current, depth and the quality of the reef and fauna that divers and snorkelers are likely to encounter.

We offer a suggested route and point out some of the key pieces of information to enhance your in-water experience, such as what species to look for and what key features to observe. When coupled with the detailed 3D renderings of the wreck or reef, you will have a good idea of what is in store before you venture into the water.

Palm Beach County has invested significant time, money and resources in developing its dive site offerings over the years, and they are still actively adding new artificial reefs when budgets and vessel availability permit. Divers and snorkelers should check with local dive shops to see what new and exciting wrecks there are to visit, and we will continue to revise and expand our guidebook in the future to help ensure we offer the most up-to-date information possible.

Diving and snorkeling services

Businesses are listed alphabetically by inlet.

Jupiter Inlet

Emerald Charters
201 Coastal Way, Jupiter
Tel: 561-248-8332
Email: randy@emeraldcharters.com
Emeraldcharters.com

Jupiter Dive Center
1001 FL A1A Alt, Jupiter
Tel: 561-745-7807
Email: info@jupiterdivecenter.com
Jupiterdivecenter.com

Jupiter Scuba Diving
1111 Love St, Jupiter
Tel: 855-575-3483
Email: bill@divekyalami.com
Jupiterscubadiving.com

Scuba Works
351 South US Hwy One, #103, Jupiter
Tel: 561-575-3483
Email: info@scubaworks.com
Scubaworks.com

Lake Worth Inlet (Palm Beach Inlet)

Brownie's Palm Beach Divers
3619 Broadway Ave, Riviera Beach
Tel: 561-844-3483
Email: info@yachtdiver.com
Yachtdiver.com/palmbeach-index.shtml

Calypso Dive Charters
200 E 13th St, Riviera Beach
Tel: 561-222-3822
Email: info@calypsodivecharters.com
Calypsodivecharters.com

Deep Obsession Charters
105 Lake Shore Dr, Lake Park
Tel: 561-360-4574
Email: dive@deepobsessioncharters.com
Deepobsessioncharters.com

Florida Scuba Charters
1037 Marina Drive, North Palm Beach
Tel: 561-676-1230
Floridascubacharters.com

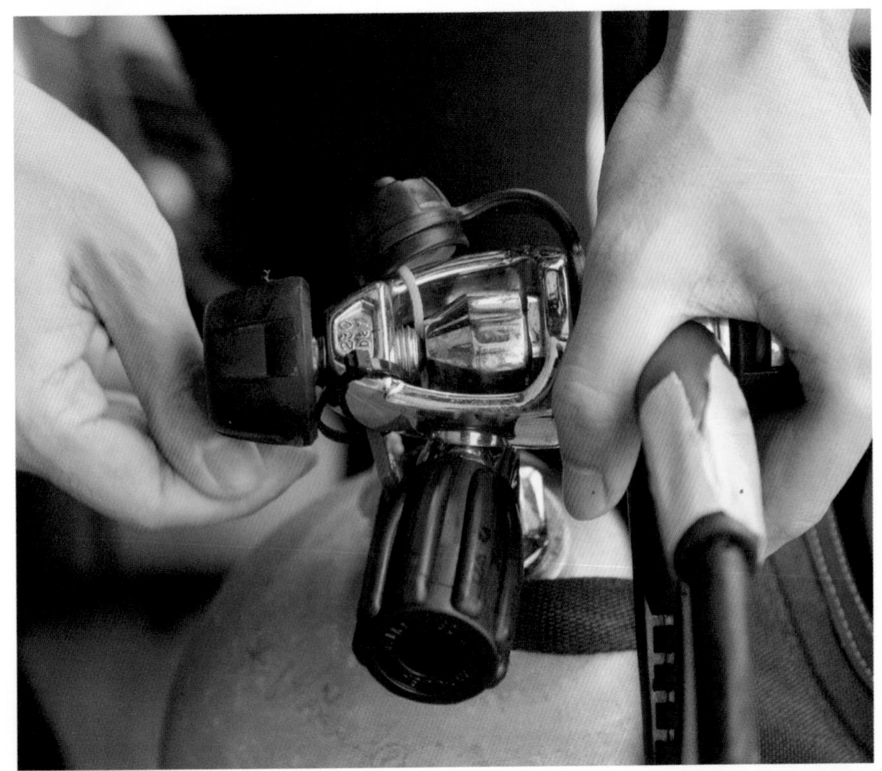

thanongsak kongtong/Shutterstock ©

DID YOU KNOW? 

Businesses close or move. They change their names or their contact information. The information on these pages is as accurate as possible at the time of printing. If you believe we have missed a shop that should be here, or if you know of information that has changed, please let us know at **info@reef-smart.com**. We will work to update the book for the next print run.

Florida Scuba Divers
635 Northlake Blvd, North Palm Beach
Tel: 561-270-5788
Email: info@floridascubadivers.com
Floridascubadivers.com

Force-E Dive Center
155 Blue Heron Blvd, Riviera Beach
Tel: 561-845-2333
Email: riviera@force-e.com
Force-e.com

Jim Abernethy's Scuba Adventures
220 North Federal Highway, Lake Park
Tel: 888-901-DIVE (3483)
Email: info@scuba-adventures.com
Scuba-adventures.com

Narcosis Dive Company
200 East 13th Street, Riviera Beach
Tel: 561-630-0606
Email: info@narcosisdive.com
Narcosisdive.com

Ocean Quest Scuba Charters
255 E 22nd Ct, Riviera Beach
Tel: 561-776-5974
Email: ocnquestscuba@aol.com
Oceanquestscuba.com

Pura Vida Divers
2513 Beach Ct, Riviera Beach
Tel: 888-348-3972
Email: info.pvd@puravidadivers.com
Puravidadivers.com
Tech.puravidadivers.com

Scuba Club
607-A Northlake Blvd, North Palm Beach
Tel: 561-844-2466
Email: info@thescubaclub.com
Thescubaclub.com

Walkers Dive Charters
200 East 13th Street, Slip P25, Riviera Beach
Tel: 561-797-DIVE (3483)
Email: bill@walkersdivecharters.com
Walkersdivecharters.com

Boynton Inlet (sometimes called South Lake Worth Inlet)

Boynton Beach Dive Center
212 S Federal Hwy, Boynton Beach
Tel: 561-732-8590
Email: boyntonbeachdivecenter@gmail.com
Boyntonbeachdivecenter.com

Force-E Dive Center
2621 North Federal Highway, Boca Raton
Tel: 561-368-0555
Email: boca@force-e.com
Force-e.com

Loggerhead Enterprises
728 Casa Loma Blvd, Boynton beach
Tel: 561-588-8686
Email: charters@loggerheadcharters.com
Loggerheadcharters.com

Splashdown Dive Center
640 E Ocean Ave #2, Boynton Beach
Tel: 561-736-0712
Email: info@splashdowndivers.com
Splashdowndivers.com

Starfish Scuba
735 Casa Loma Blvd Slip #11, Boynton Beach
Tel: 561-212-2954
Email: info@starfishscuba.com
Starfishscuba.com

Underwater Explorers
735 Casa Loma Blvd, Boynton Beach
Tel: 561-577-3326
Email: kevin@diveboyntonbeach.com
Diveboyntonbeach.com

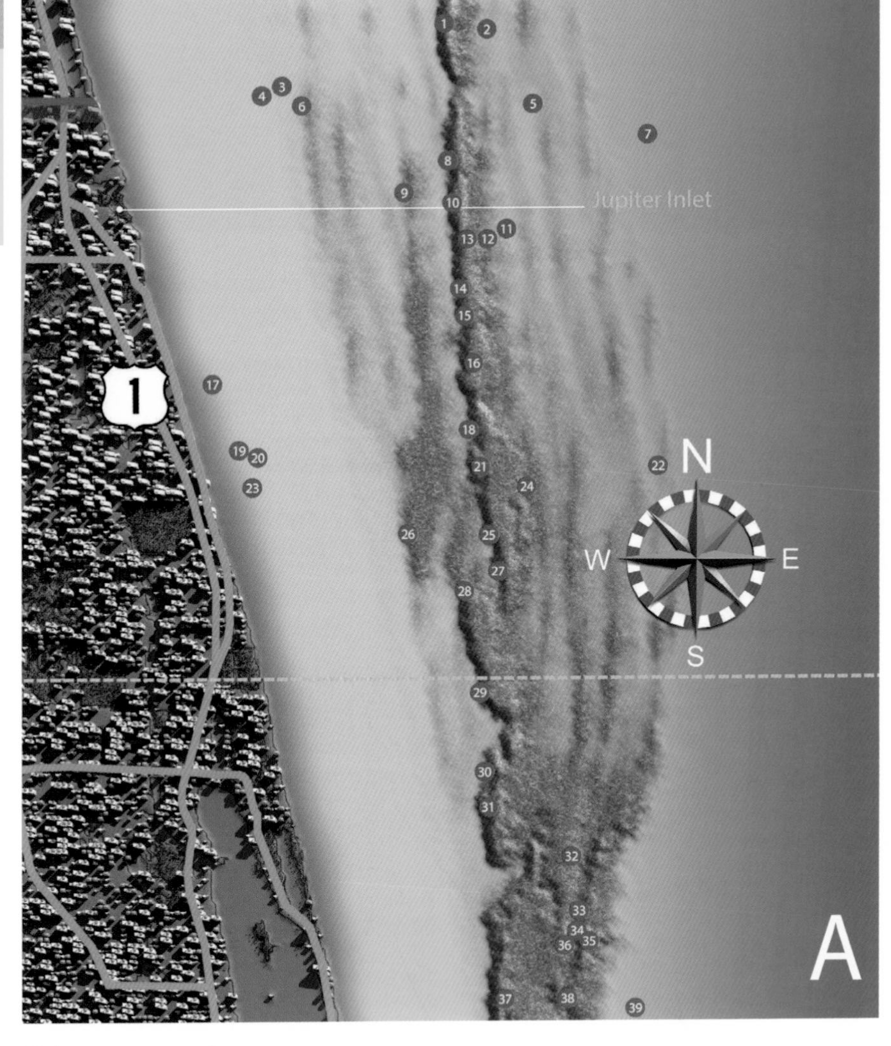

Dive and Snorkel Sites

1 *MG-111* and Warrior Reef
2 Tug Boat Reef
3 Andrew Red Harris Reef
4 Jupiter Stepping Stone Reef
5 Jupiter Wreck Trek
6 Andrew Harris No Shoes Reef
7 *Sea Mist II*
8 Julies
9 Tunnels (Inner Reef)
10 Loggerhead (Jupiter Ledge)
11 *Zion Train* Bow
12 Red Grouper Hole
13 Lighthouse
14 Bonnie's (need location)
15 Captain Kirle's
16 Scarface
17 Diamonhead Radner Reef
18 Bluffs
19 Kwiecinski Reef
20 Juno Pier North
21 Spadefish
22 Hole in the Wall
23 Juno Pier South
24 Area 29
25 Area 51
26 Andrew Red Harris Foundation Juno
27 Captain Mike's
28 Juno Ledge
29 Shark Canyon
30 26 51 Ledge
31 26 50 Ledge
32 26 49 Ledge
33 Zeigler's Zoo
34 Condos (100' Caves)
35 Calloway's Ledge (Black Rock)
36 Mystery Wreck
37 *Runaway Barge* (*Murphy's Barge II*)
38 Mystery Wreck
39 Vance's Reef

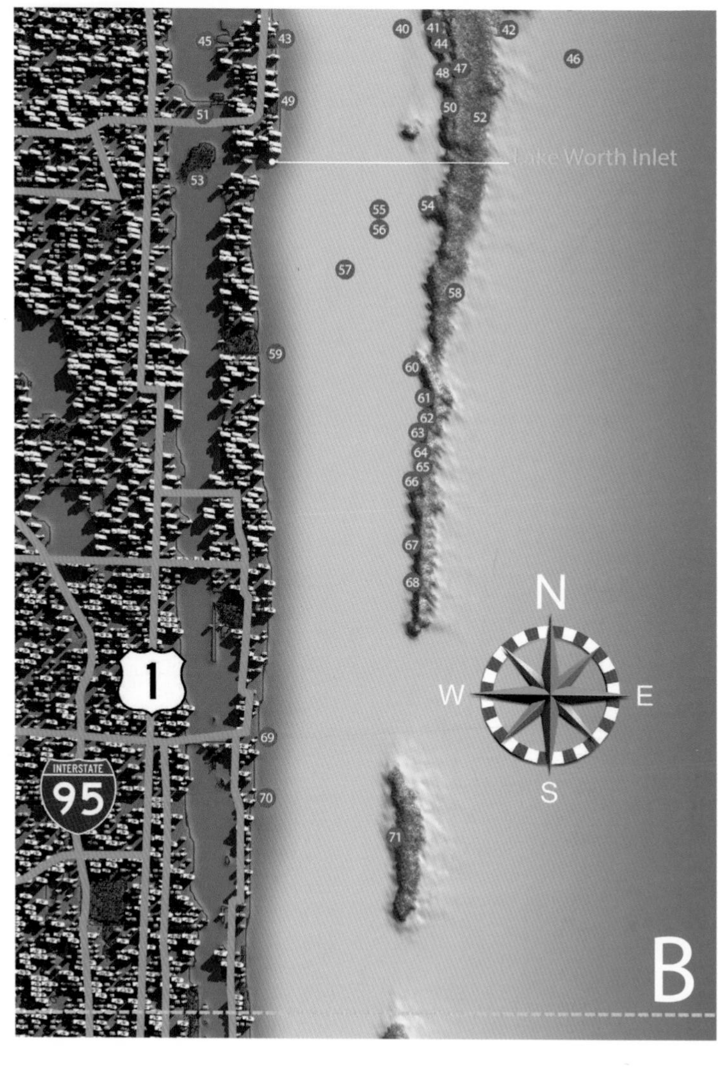

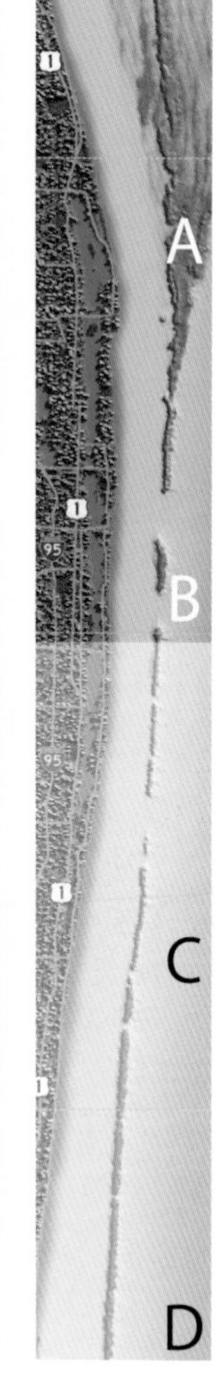

40 Royal Park Bridge Reef
41 Danny McCauley Memorial Reef & Atlantis
42 Princess Ann Reef
43 Ocean Reef Park
44 Brazilian Docks
45 Sugar Sands Ledges
46 Ande Fishing Reef
47 *Simpson's Barge*
48 Mizpah Corridor Wreck Trek
49 Singer Island Reef
50 *Ana Cecilia*
51 Blue Heron Bridge/ Phil Foster Snorkel Reef
52 *Spearman's Barge*
53 Peanut Island Reefs
54 Eidsvag Triangle
55 Tri-County Reef and TSO Paradise (Playpen)
56 Cross Current Reef
57 Flagler Bridge Reef PB Reefs (three reefs)
58 Governor's River Walk Reef
59 Breakers Shallow
60 Breakers-Turtle Mounds (Lopata's Reef)
61 Fourth Window
62 Trench
63 Breakers South
64 Flower Gardens
65 East-West Ledge (Middle Earth)
66 Teardrop
67 Bath and Tennis Reef
68 Cardinal's Reef
69 Mar-A-Lago Reefs
70 Widener's Curve Mitigation Reef
71 Paul's Reef

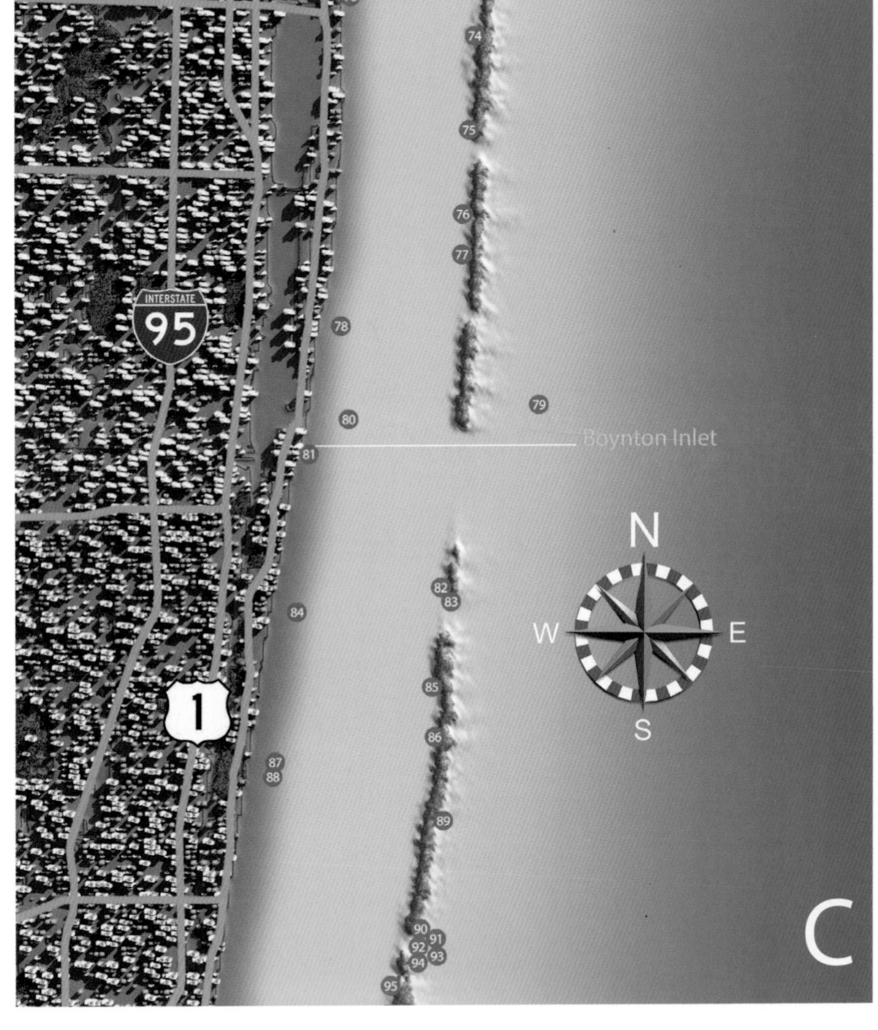

72 Palm Beach Horsehoe

73 Kreusler Park
74 Fay's Reef
75 Lady Anne Reef
76 Big Brain Reef
77 Fish Bowl
78 *SV Lofthus*
79 Boynton Kiwanis Miller Lite Reef (Skycliff)

80 Boynton Stepping Stone Reef
81 Ocean Inlet Park Snorkel Reef (Boynton Inlet)
82 Donny Boy Silpe Reef
83 Ragg's Reef
84 Boynton Beach Ocean Park Reef
85 Lynn's Reef
86 Briny Breezes Ledge
87 Gulfstream Mitigation Reef North

88 Gulfstream Mitigation Reef South
89 Boynton Ledge

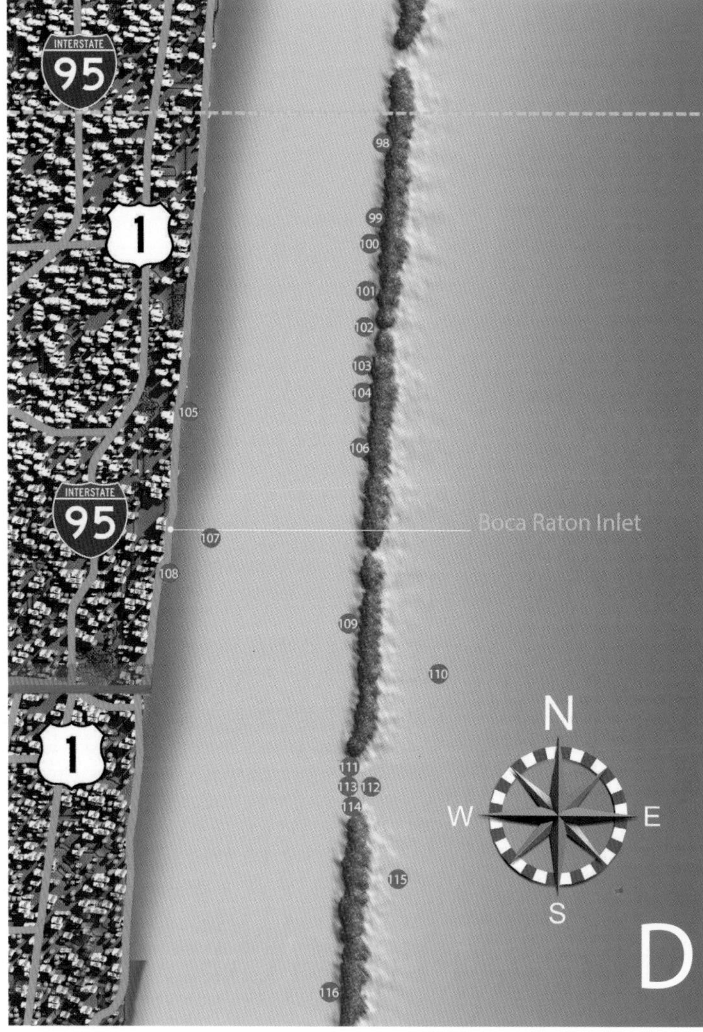

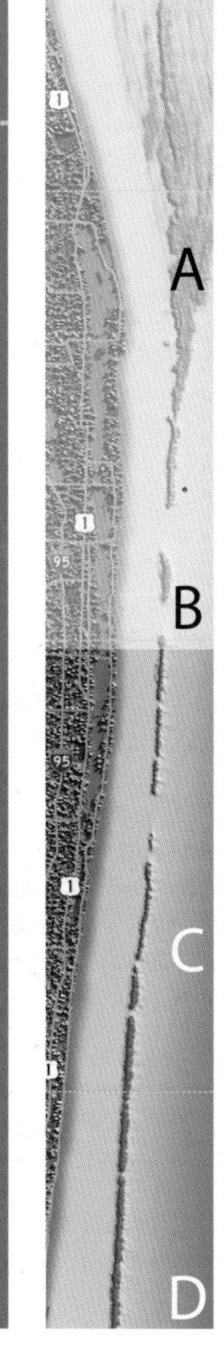

90	*MV Becks* (*Captain Tony*)	106	Boca Third Reef Ledge
91	Boynton Corridors North	107	Boca Step Reef
92	*Budweiser Bar* (*MS Havel*)	108	South Boca Inlet Reef
93	*MV Castor*	109	Boca Artificial Reef Ledge
94	Boynton Corridors South	110	*Hydro Atlantic*
95	Delray Ledge (Hobie Ledge)	111	*Sea Emperor*
96	Seagate Grouper Hole	112	*Noulla Express*
97	*SS Inchulva*	113	Boca Corridors
98	Sally's Reef	114	*United Caribbean*
99	Fink's Grouper Hole	115	*Miss Lourdies*
100	Grouper Hole South	116	*Ancient Mariner*
101	Highland Beach Reef		
102	Yamato Rock 3rd Reef		
103	Sand Chute		
104	Boca North Beach Ledge		
105	Red Reef Park		

MG-111 & Warrior Reef

Difficulty ●○○
Current ●●○
Depth ●●○
Reef ★☆☆
Fauna ★★☆

Access about 25 mins from Jupiter Inlet

Level Open Water

Location

Jupiter, Palm Beach County
GPS (*MG-111*): 26°58'38.2"N, 80°01'29.6"W

Getting there

MG-111 and Warrior Reef represent a two-part artificial reef just over 3 miles (5 kilometers) off shore of Jupiter – about a 25-minute boat ride northeast of the inlet. The best way to reach *MG-111* and Warrior Reef is through one of the area's local dive operators using the Jupiter Inlet.

Access

There is no permanent mooring buoy on *MG-111*, so local operators will often put divers in the water slightly up current from the site so they have time to descend to the seafloor and drift onto the site. Visibility is generally good on this site, but currents can be strong. This dive is accessible to divers of all experience levels as it is relatively shallow and there are no penetration opportunities on *MG-111*.

Description

MG-111 is a Mississippi river barge deliberately sunk as an artificial reef in 1995. The barge is covered in a jumbled pile of concrete poles once used for street lighting. The 1,000 tons of debris helps provide habitat for reef fishes that have colonized the wreck. The 195-foot (59-meter) barge is largely unremarkable on its own. It has collapsed under the weight of the concrete beams, leaving just the bow and stern visible as two triangular sections at either end of the wreck. It is the diversity of marine life present on this artificial reef that brings divers back time and again. In particular, divers regularly encounter solitary goliath grouper near the barge or on the adjacent Warrior Reef.

During the breeding season, divers may see the grouper in an active spawning aggregation, swimming together in a large group and releasing gametes (eggs and sperm) into the surrounding waters, visible as a milky white cloud. Swimming carefully with slow movements and controlled breathing is usually required to approach grouper, but during spawning they are more focused on each other than on the divers nearby.

Just 50 feet (15 meters) north of the wrecked barge stands a field of concrete pillars known as Warrior Reef. Deployed in 2005, the 250 tons of concrete pillars were once part of an outdoor,

SCIENTIFIC INSIGHT

MG-111 is a goliath grouper spawning site – one of several along the coast of Palm Beach County and the closest known site to shore. During their spawning season, which occurs from July to September, goliath grouper from up to 350 miles (563 kilometers) away gather at a handful of sites. They come together to spawn in groups of as few as 10 to as many as 100 – typically coinciding with the new moon.

Scientific research suggests that many fish species form spawning aggregations, similar to goliath grouper, at distinct locations, which are often places where currents are strong, such as pinnacles, channels between reefs and on outer reef slopes. The reason for this choice in sites is that the currents help take the gametes (eggs and sperm) away from the reef, which increases their chance of survival.

Commercially important species are extremely vulnerable to over exploitation if spawning sites are targeted by fishers. For instance, Nassau grouper populations across the Caribbean have been devastated for this reason, and are now considered critically endangered – one step away from extinction. Thankfully, these days more grouper spawning sites are protected than ever before, which will hopefully help the species recover.

covered walkway located on the campus of a local high school. The 10-foot-tall (3-meter) pillars were donated by the Jupiter High School (the home of the Warriors sports team, hence the name). They were deployed in upright positions across the sandy bottom to the north of the barge. Some of the pillars have fallen down since their initial deployment, but most remain upright albeit with a slight tilt.

Together, *MG-111* and Warrior Reef provide habitat for a variety of reef fish species, including schools of porkfish, grey snapper, sergeant majors and tomtates. Stingrays are also regular visitors and can be found cruising the sand near the wreck and the adjacent columns.

The main attraction in terms of fauna, namely the goliath grouper, are most often seen stacked one on top of each other next to the columns of Warrior Reef. Divers can carefully approach the resting giants from down near the sand, often getting within just a few feet of the grouper as they swim in place. During spawning season, these giant fish do not startle easily, so chances of approaching relatively closely are much higher than at other times of the year.

Goliath grouper shelter from the current behind a pillar on Warrior Reef.

Jupiter Dive Center ©

Route

The currents along this part of the Florida coastline typically run from south to north. Most dives are conducted as drift dives, starting with the *MG-111* before heading north into the field of concrete pillars that comprise Warrior Reef.

Depending on the strength of the current, divers may be able to complete a full circuit of the barge before continuing on to the columns. If currents are particularly strong, however, it may only be possible to cover one side of the wreck.

Divers should check the covered areas under the triangular-shaped sections of wreckage that are the remnants of the bow and stern of the barge for solitary grouper. Large fish often shelter in the shadowy recesses. When divers are done touring the barge, they can make their way north toward the columns of Warrior Reef.

When goliath grouper are present in Warrior Reef, divers should enter the field of pillars slowly, controlling their breathing to avoid startling the large fish – this is particularly important when diving with a large group. Since the sandy seafloor is only 65 feet (20 meters) deep, divers should have plenty of bottom time to spend watching the grouper.

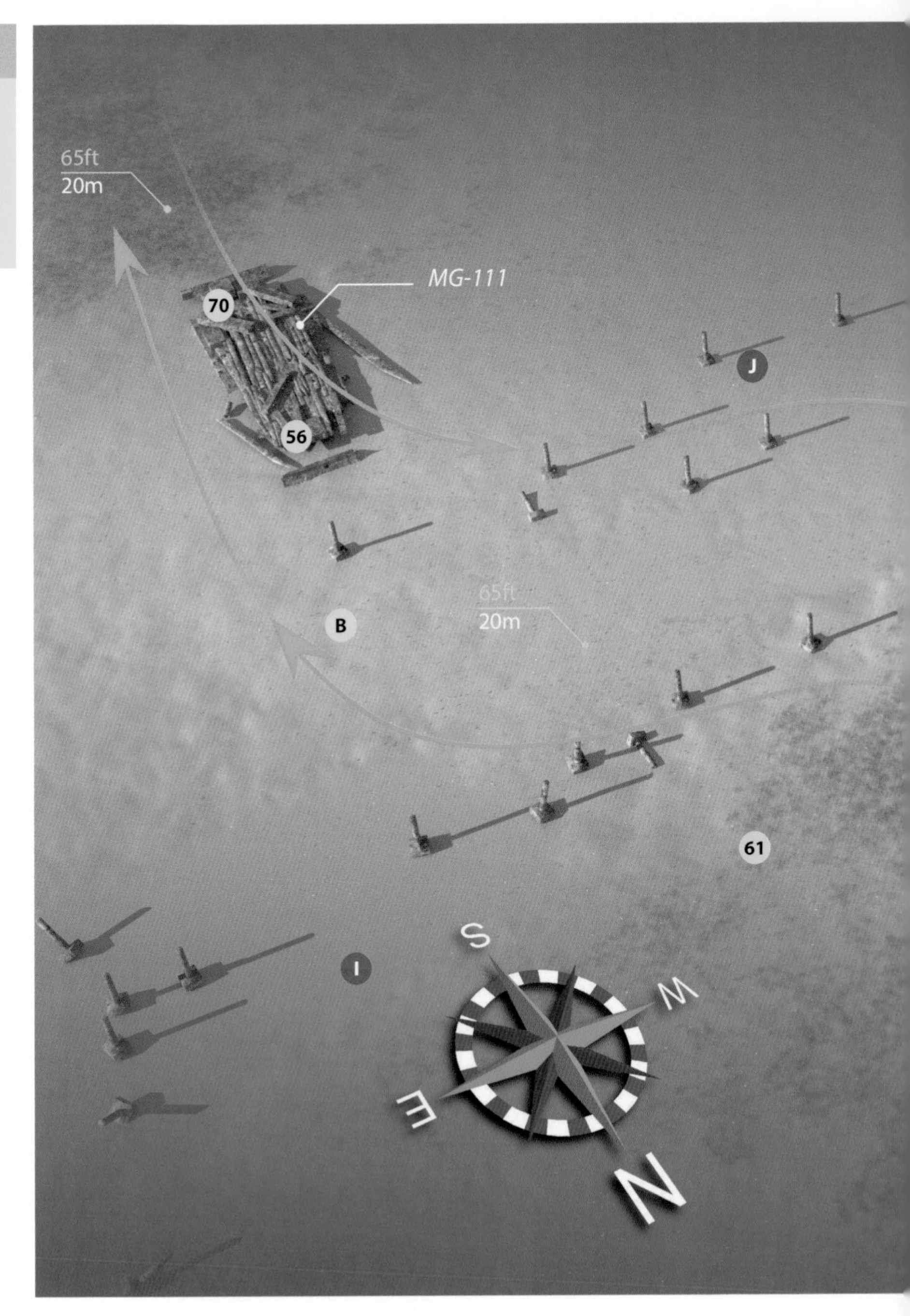
65ft
20m
MG-111
70
56
J
B
65ft
20m
61
I
S
W
E
N

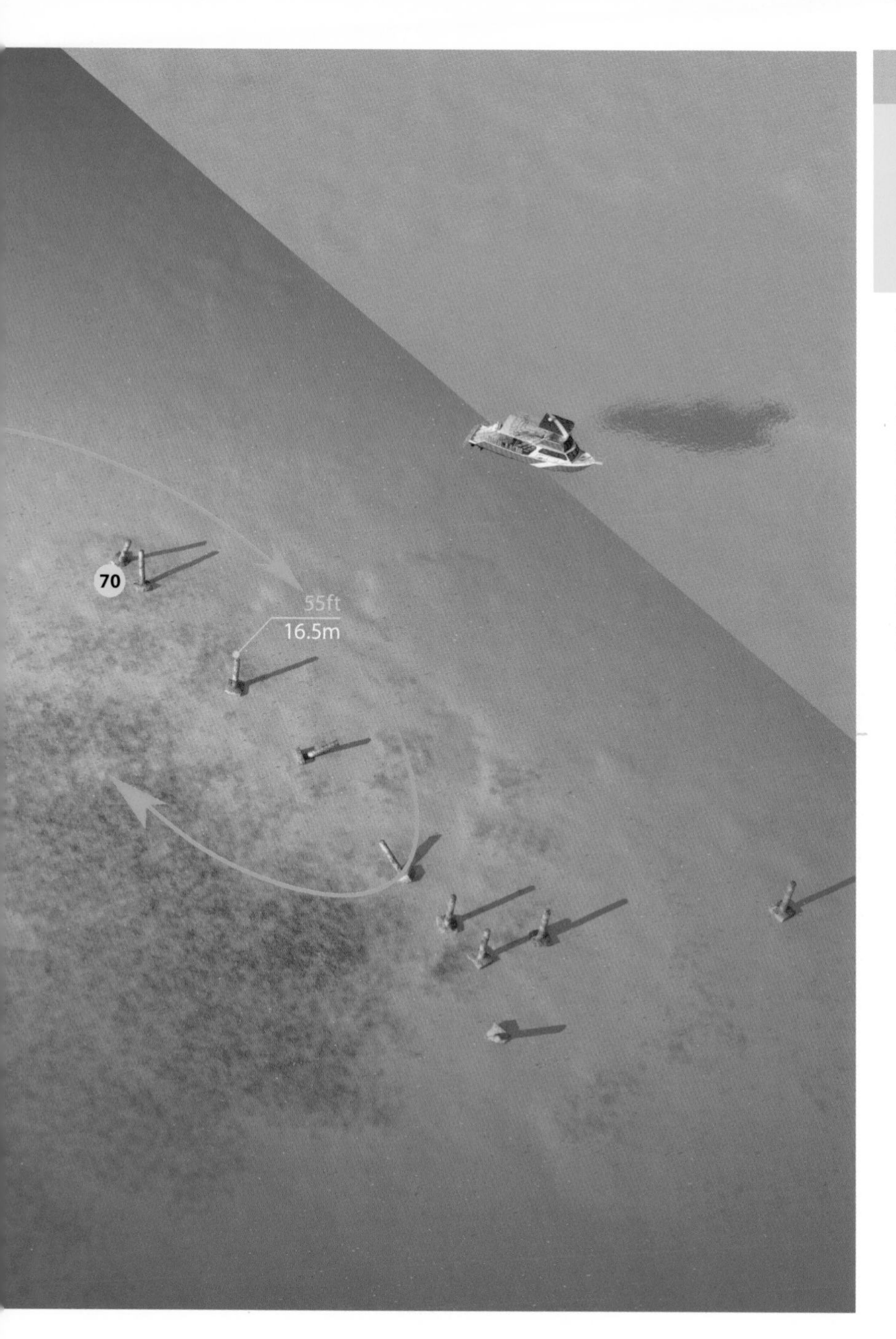

Other species commonly found at this site: L 8 11 13 15 18 19 20 27 32 36 38 44 45 52 67

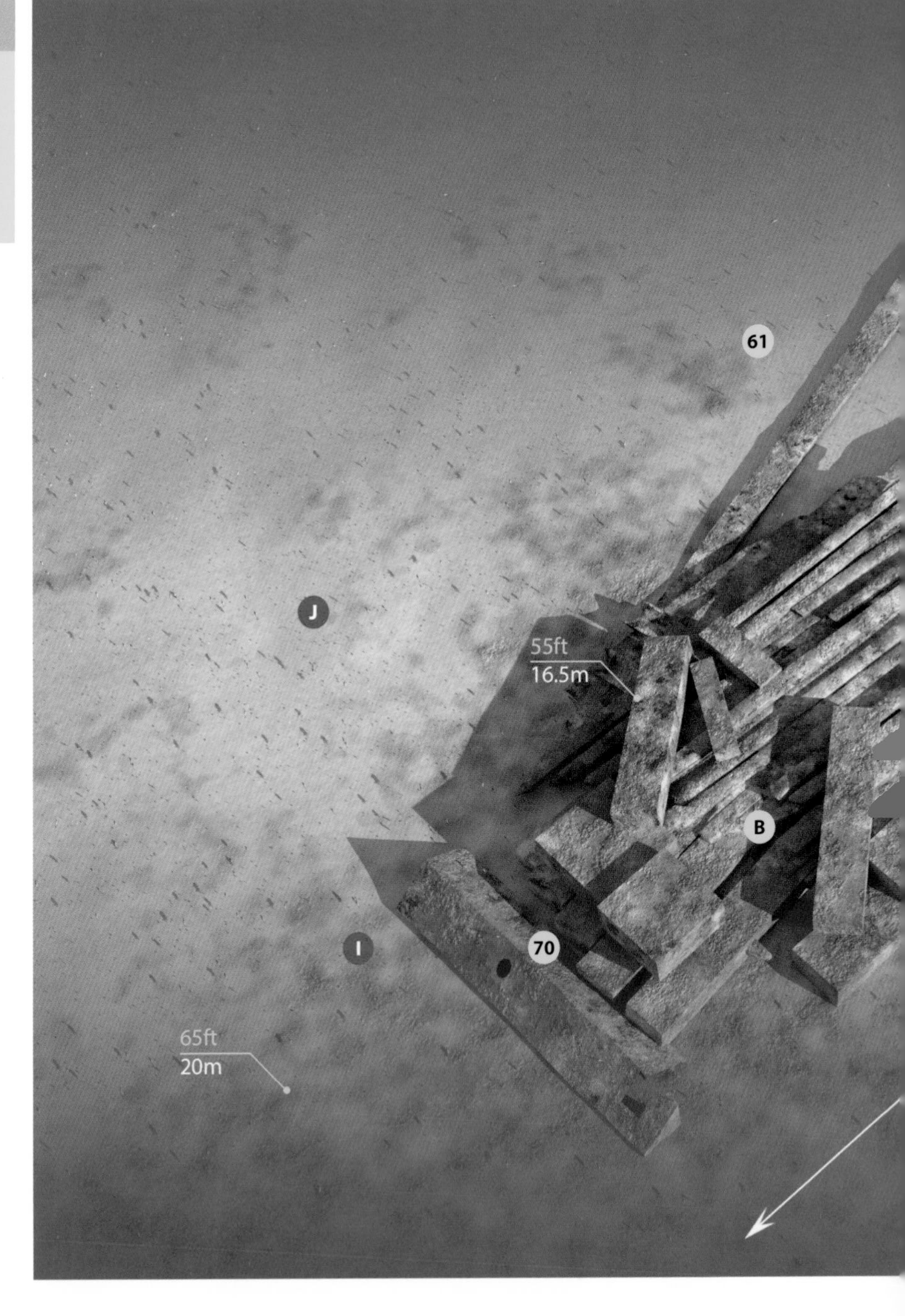

Name:	*MG-111*	**Construction:**	Unknown
Type:	River barge	**Last owner:**	Unknown
Previous names:	n/a	**Sunk:**	September 28, 1995
Length:	195ft (58m)		
Tonnage:	Unknown		

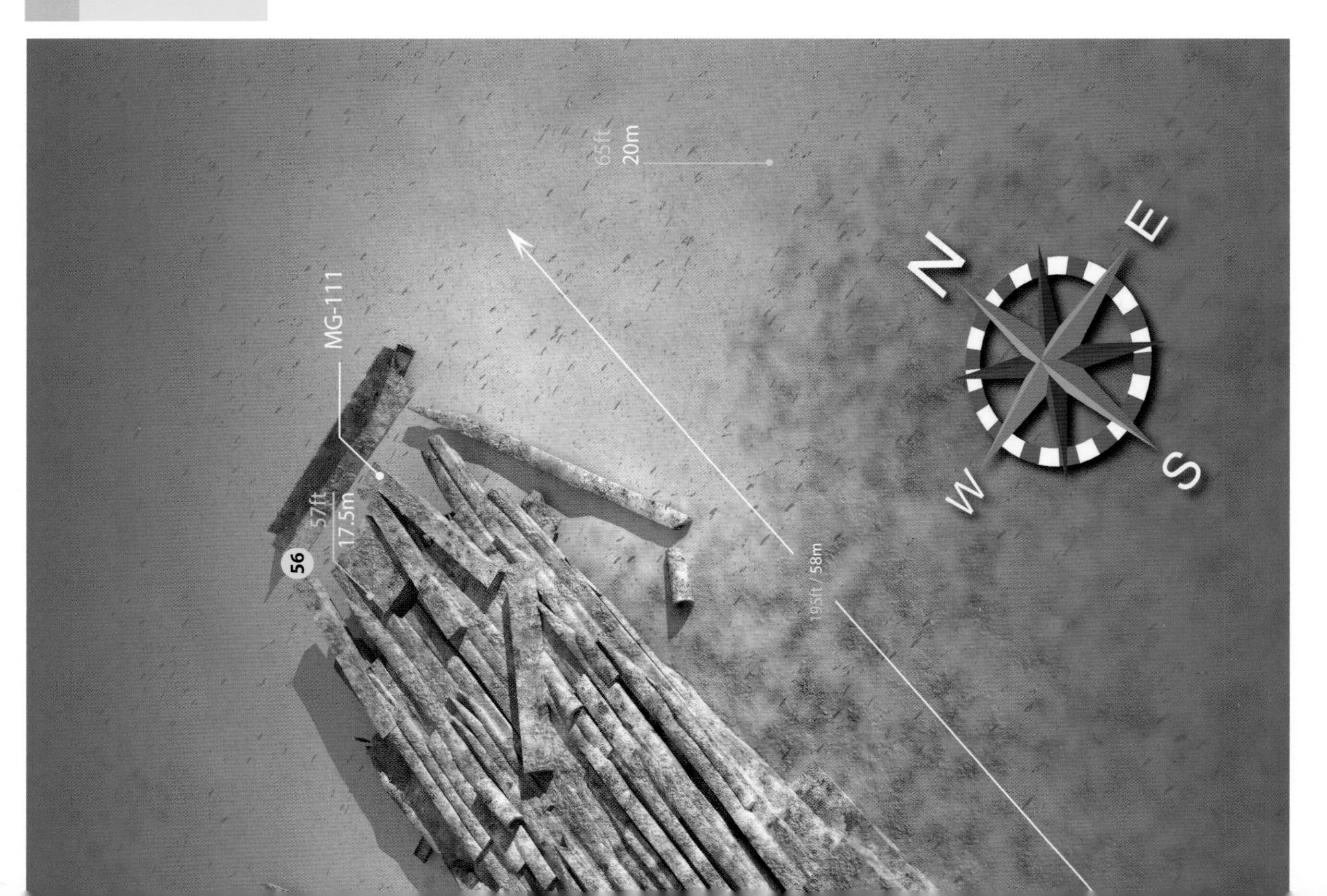
65ft
20m
MG-111
57ft
17.5m
56
195ft / 58m
N
E
W
S

Jupiter Wreck Trek

Difficulty ●●○
Current ●●○
Depth ●●○
Reef ★☆☆
Fauna ★★☆

Access about 30 mins from Jupiter Inlet

Level Advanced Open Water

Location

Jupiter, Palm Beach County
GPS (*Zion Train*): 26°57′47.2″N, 80°00′26.9″W

Getting there

The Jupiter Wreck Trek consists of three wrecks sunk in a line running north to south, just over 4 miles (6.5 kilometers) off shore from Jupiter – about a 30-minute boat ride east of the inlet. The best way to reach the wreck trek is through one of the dive operators located in and around the Jupiter Inlet for ease of access.

Access

There is no permanent mooring buoy on any of the wrecks in this trek. Local operators often put divers in the water slightly up current of the site so they can descend toward the bottom and allow the currents to push them to the wrecks. Divers typically explore the trek as a drift dive from south to north, following the current. Visibility is generally good on this site, but currents can be strong. This dive is more suitable for divers with an advanced open water certification as the seabed bottoms out at an average of 90 feet (27.5 meters) and the wrecks offer limited penetration. Diving on nitrox is recommended to provide the bottom time necessary for divers to fully explore the wrecks.

Description

The Jupiter Wreck Trek includes three wrecks, the stern section of the *Zion Train*, the *Miss Jenny* and the *Esso Bonaire III*. (The bow section of the *Zion Train* is not included in the trek as it

A diver gets picked up after a drift dive on the Jupiter Wreck Trek.

is currently located 1.5 miles (2.4 kilometers) to the southwest of the stern.) Some divers and dive shops also refer to the entire wreck trek by the name *Zion Train*. Either way, it is popular with divers seeking to encounter mega fauna such as goliath grouper and sharks, including reef, lemon and even bull sharks.

The *Zion Train* was originally a 164-foot (50-meter) coastal freighter built in Holland in 1962. She started out as a cargo vessel in European waters, changing ownership and names a few times before crossing over to the Caribbean and ending up with a Honduran flag and a Haitian crew in the late 1990s. As the story goes, the ship was in port on the Miami River when alleged pirates boarded and shot many of the crew members. Just a few months later, the ship ran aground near South Beach, Florida, before being seized by authorities. It was eventually purchased by Palm Beach County and deliberately sunk as an artificial reef in June of 2003.

The stern and bow sections separated when Hurricanes Frances and Jeanne passed through the region in 2004. The back-to-back storms managed to push the bow section more than a mile away, and well beyond the scope of divers exploring the trek as a drift dive. The stern section sits listing on its starboard side, facing south. The hull sections also lie flattened and strewn to the south and east of the stern. The structure does not boast much in the way of coral growth, but it supports many large goliath grouper. Stingrays are often spotted in the sand around the wreck.
Located just 200 feet (61 meters) to the north of the *Zion Train*'s stern section is the dredge known as *Miss Jenny*. The small vessel was sunk nearby to a pile of concrete pilings named "concrete light poles" by the county's artificial reef program. The poles lie on the sandy seafloor to the east of dredge. The roughly square-shaped dredge offers little in the way of coral growth despite having been underwater for nearly three decades, but it often attracts goliath grouper during spawning season and is a handy navigational marker. The county sank the dredge as an artificial reef in 1990.

The wreck of the *Esso Bonaire III* sits 500 feet (152 meters) to the north-northwest of *Miss Jenny*. She completes the three-wreck trek, and is the most interesting, structurally speaking, of the three vessels. The county's artificial reef program sank *Esso Bonaire III* in July 1989 after authorities discovered 55,000 pounds of marijuana onboard during a search. The Honduran-flagged vessel was impounded and slated for sinking on July 22,

Jupiter Dive Center ©

ECO TIP

Some dive operators believe feeding sharks can enhance the diver experience. This practice is not allowed within Florida's state waters, which extend out to 3 miles (5 kilometers). But since the Jupiter wreck trek sits outside of that limit, dive operators can legally attract sharks with chum. Even so, this practice is generally discouraged by most biologists.

For one, sharks have been known to get more aggressive where feeding occurs, which can result in injury to divers. Moreover, sharks habituated to feeding at a chum station when divers are around are no longer exhibiting natural behavior. An up-close and personal encounter with a shark can be a thrilling experience. But we strongly encourage you to avoid the temptation of seeking out an artificial, and potentially dangerous, encounter created through chumming unless it is part of a well-organized educational experience.

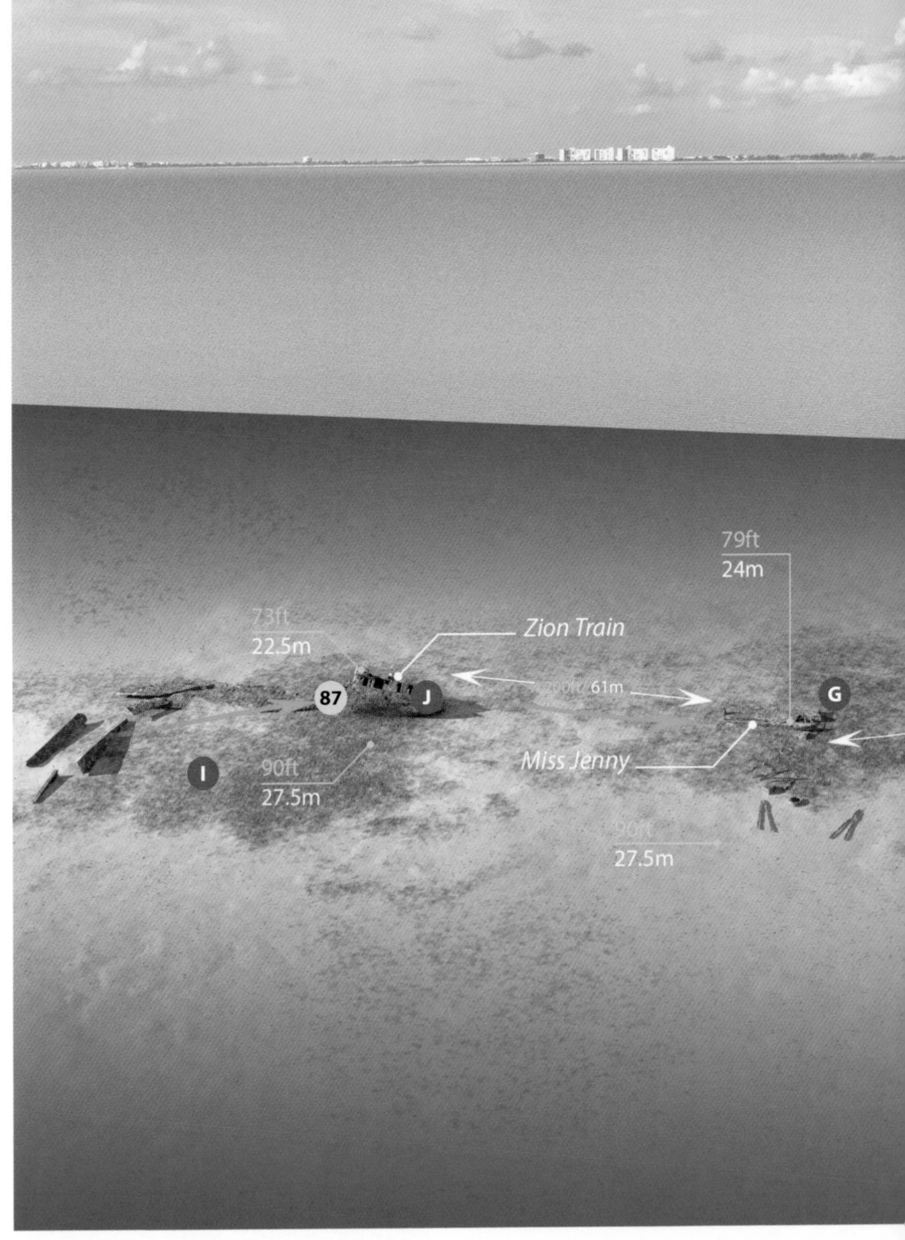

but bad weather delayed this by a day. The wreck allegedly took more than nine hours to finally slip beneath the waves.

The wreck currently sits upright on the sandy seafloor at a depth of 90 feet (27.5 meters). She is relatively intact and heavily encrusted with gorgonians. Unfortunately, a boat that was tied off to the wreck during strong currents ended up damaging a section of the wreck's superstructure. This damage left a twisted section of the bridge partially blocking one of the swim-through opportunities. Despite the damage, the structure of the wreck remains in good condition and attracts plenty of goliath grouper that hide out in the dark recesses of the wreck. Lucky divers may also have the chance to see spotted eagle rays, cobia and barracuda at this site.

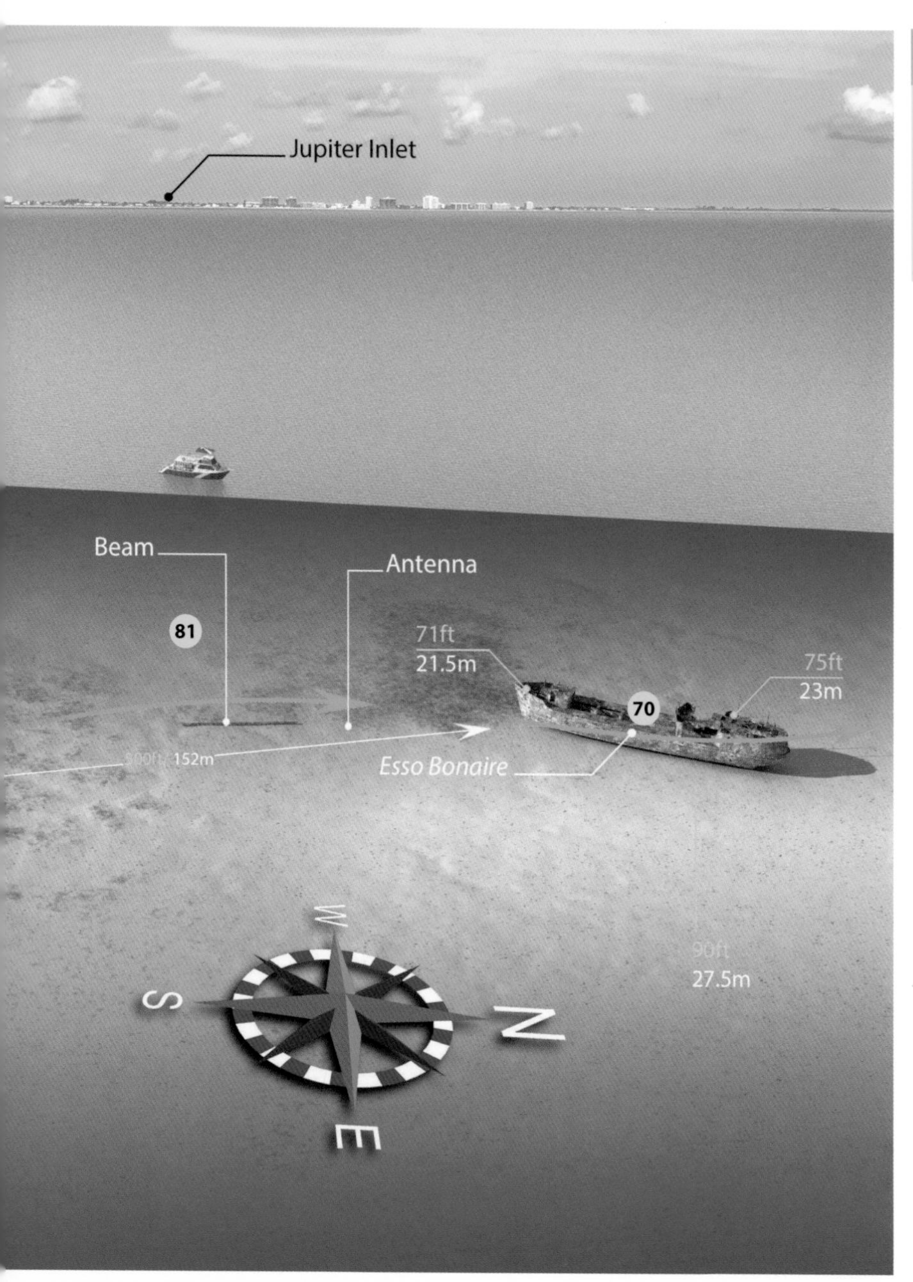

Just off the stern of the *Esso Bonaire III*, and just beyond a length of concrete lying perpendicular to the wreck's orientation, some operators chum the waters to generate up-close interactions

Other species commonly found at this site: N 2 22 25 42 52 56 65 66 69 71 72 74 75 77 79

Zion Train position

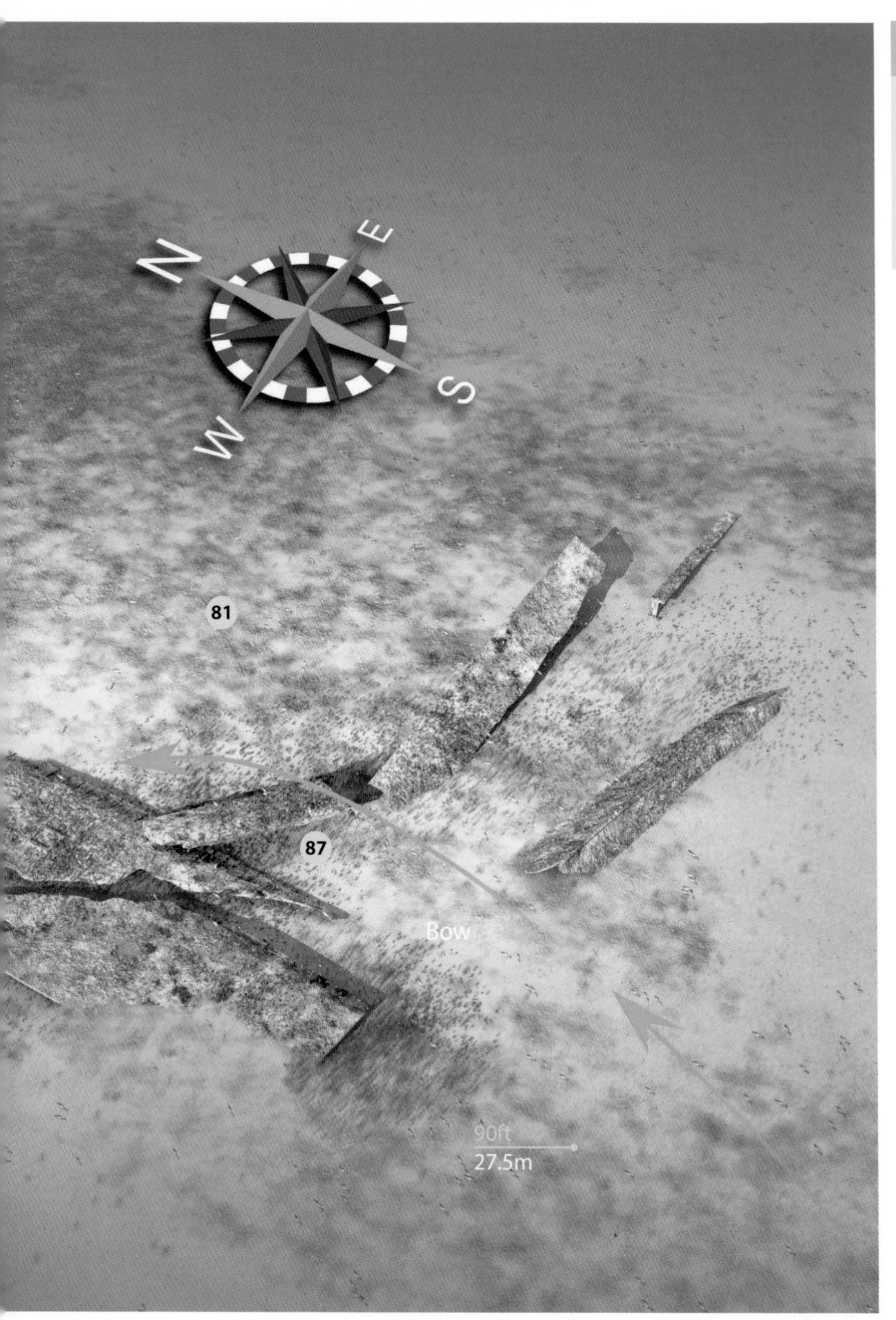

Name:	*Zion Train*	**Last owner:**	Association Industriale Haitien, San Lorenzo
Type:	Freighter		
Previous names:	*MV Plancius, MV Novel*	**Sunk:**	June 6, 2003
Length:	164ft (50m)		
Tonnage:	399grt		
Construction:	Van Der Werff G. J. Scheepsbouw, Westerbroek, 1962		

J
90ft
27.5m
G
E
S
N
W
I
Miss Jenny position
Esso Bonaire
N
W
E
S

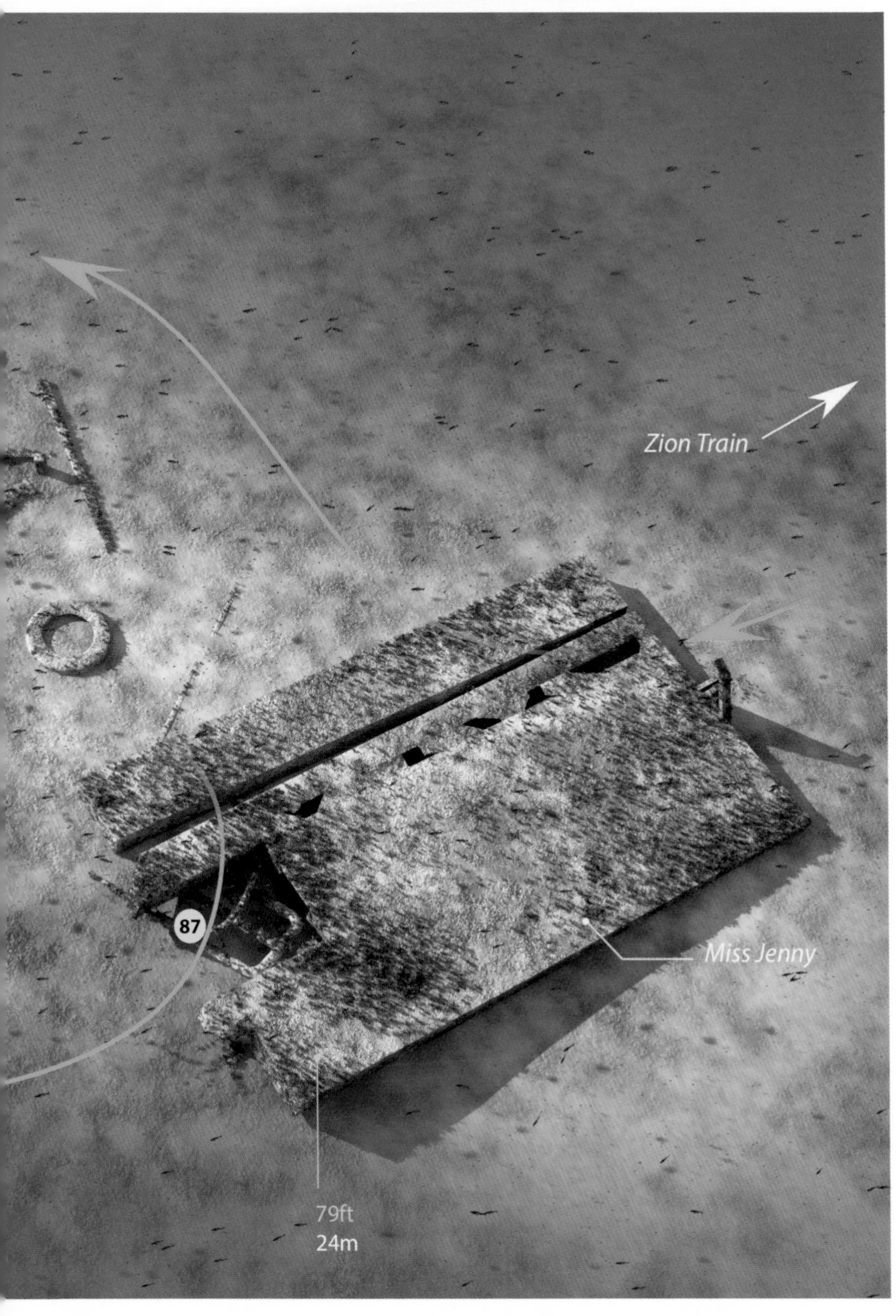

Name:	*Miss Jenny*	**Last owner:**	Unknown
Type:	Dredge	**Sunk:**	December 11, 1990
Previous names:	n/a		
Length:	55ft (17m)		
Tonnage:	100grt		
Construction:	Unknown		

with sharks. Due to the repeated chumming, it is common for sharks to cruise through this area when divers are present.

Route

Divers typically start out approaching the *Zion Train* from the south. An anchor chain leads into the debris field near the stern section of the wreck. Divers often only spend a few minutes on the *Zion Train* – more if there are goliath grouper hanging out near the wreck. From there, divers drift due north along a path marked by rebar in the seafloor toward *Miss Jenny*. Goliath grouper often hang out near the dredge as well, particularly during the summer months. Divers typically spend only a brief moment on the dredge, primarily using it as a navigational reference on their way to the *Esso Bonaire III* – the star of the trek.

The *Esso Bonaire III* sits 500 feet (152 meters) along a compass heading of 340 degrees from the dredge. Again, rebar marks this route, and if divers drift too far to the west, they will encounter a long, concrete beam partly buried in the sand, which lines up with an upright antenna-like structure. The *Esso Bonaire III* sits due north of this antenna.

Divers often spend the bulk of their dive time exploring the *Esso Bonaire III*, usually completing a full circuit of the 147-foot-long (45-meter) freighter if currents allow. Divers can access the holds of the wreck to look for resting grouper. As the dive comes to an end, most divers drift off the back of the ship as they ascend, which may provide a glimpse of lemon sharks as they swim through the sandy patch to the north of the wreck.

Name:	*MV Esso Bonaire III*	**Last owner:**	Resolve Salvage and Towing
Type:	Tanker	**Sunk:**	July 23, 1989
Previous names:	n/a		
Length:	147ft (45m)		
Tonnage:	402grt		
Construction:	Camden Yards, NJ, 1939		

71ft
21.5m
J
Bow
90ft
27.5m
70
81
S
E
W
N
Esso Bonaire position
Esso Bonaire
Antenna
Beam
N
W
E
S

Tunnels

Difficulty ●●○
Current ●●●
Depth ●●○
Reef ★☆☆
Fauna ★★★

Access about 16 mins from Jupiter Inlet

Level Open Water

Location

Jupiter, Palm Beach County
GPS: 26°56′45.0″N, 80°01′52.2″W

Getting there

Tunnels is located on a relatively short section of reef ledge that runs parallel to the coast just 2.5 miles (4 kilometers) off shore from the Jupiter Inlet. It is only accessible by boat due to its distance from shore. The best way to reach it is with a local dive operator that uses this inlet.

Access

This site is accessible to divers of all experience levels. Due to its north-south orientation, the ledge lies parallel to the prevailing currents. Most divers explore this site as a drift dive. Visibility is generally good. Currents can be strong, which helps divers explore more ledge on their dive. There are no mooring buoys along this ledge so divers should use a surface marker buoy.

Description

Tunnels is a dive site that rarely disappoints. It gets its names from the many swim-throughs that divers can explore as they drift along the face of the ledge. But it is best known for the mega fauna regularly spotted here, such as reef sharks, goliath grouper and sea turtles, including green, hawksbill and loggerheads. Divers who can remember to tear their eyes away from the ledge to look out across the sand and rubble bottom to the west will likely also see large southern stingrays hunting for food or settling down to rest on the seafloor.

The ledge at Tunnels bottoms out at a depth of around 75 feet (23 meters), which means there is plenty of time to explore whether on air or nitrox, although nitrox divers may get to explore some of the interesting features at the northern extreme of the site if currents allow.

The top of the ledge at Tunnels runs at a fairly consistent depth of 65 feet (20 meters) although it drops to 70 feet (21.5 meters) in a few places. It has very little coral cover, however, with only a few sponges and gorgonians to break up the carpet of macroalgae and turf algae that cover the substrate. For this reason, most divers spend their time just off the ledge exploring the swim-throughs, boulders and undercuts along the interface with the sand. One of the most interesting areas at this site is located at the start of the dive, where part of the ledge extends out toward the sand in a narrow horseshoe shape, creating a swim-through

Schooling Atlantic spadefish are often found at Tunnels.

RELAX & RECHARGE

U-Tiki Beach is nestled alongside a pretty marina in Jupiter with an outstanding view. You can even take a boat ride around the marina to view the nearby lighthouse. There is a kids' menu; a beach menu with snacks, including sushi rolls; a lunch menu, featuring burgers, salads, wraps and flatbreads; and finally a dinner menu that includes Caribbean dishes and specialty plates, such as filet mignon, hogfish and mahi mahi. U-Tiki does not take reservations, so get there early – put your name on the list and enjoy a sundown drink while you wait for your table. It can take some time, but you will not be disappointed – there is a good reason why this place is so busy. Visit: **Utikibeach.com**

Peter Leahy/Shutterstock ©

between two recessed sections of the ledge. Goliath grouper are commonly spotted hanging out in the protection of this first swim-through.

Multiple undercut areas and large boulders lie to the north of the first swim-through, where the ledge has collapsed. These areas offer divers the chance to look for lobster hiding in the crevices. Schools of grey snapper are dense enough to almost hide the ledge entirely in some places, while large dog snapper cruise along the face of the ledge, alongside angelfish and parrotfish, including the rare blue parrotfish. However, none of the swim-throughs to the north offer the same level of structure and complexity as the first one.

Divers often encounter large schools of Atlantic spadefish as they drift along the ledge. They may also see sea turtles grazing along the top of the reef, while barracuda patrol in the water column above. At the far northern end of the site is an area known as the "donut hole," where divers can

swim through the ledge wall, up a chimney (the hole in the donut hole) and out onto the top of the reef. Reef sharks are a regular sight along this section of the ledge.

Route

Most divers drift along the ledge from south to north with the prevailing currents. Divers are often dropped a few minutes south of the first swim-through, but exact placement will depend on the strength of the current and the

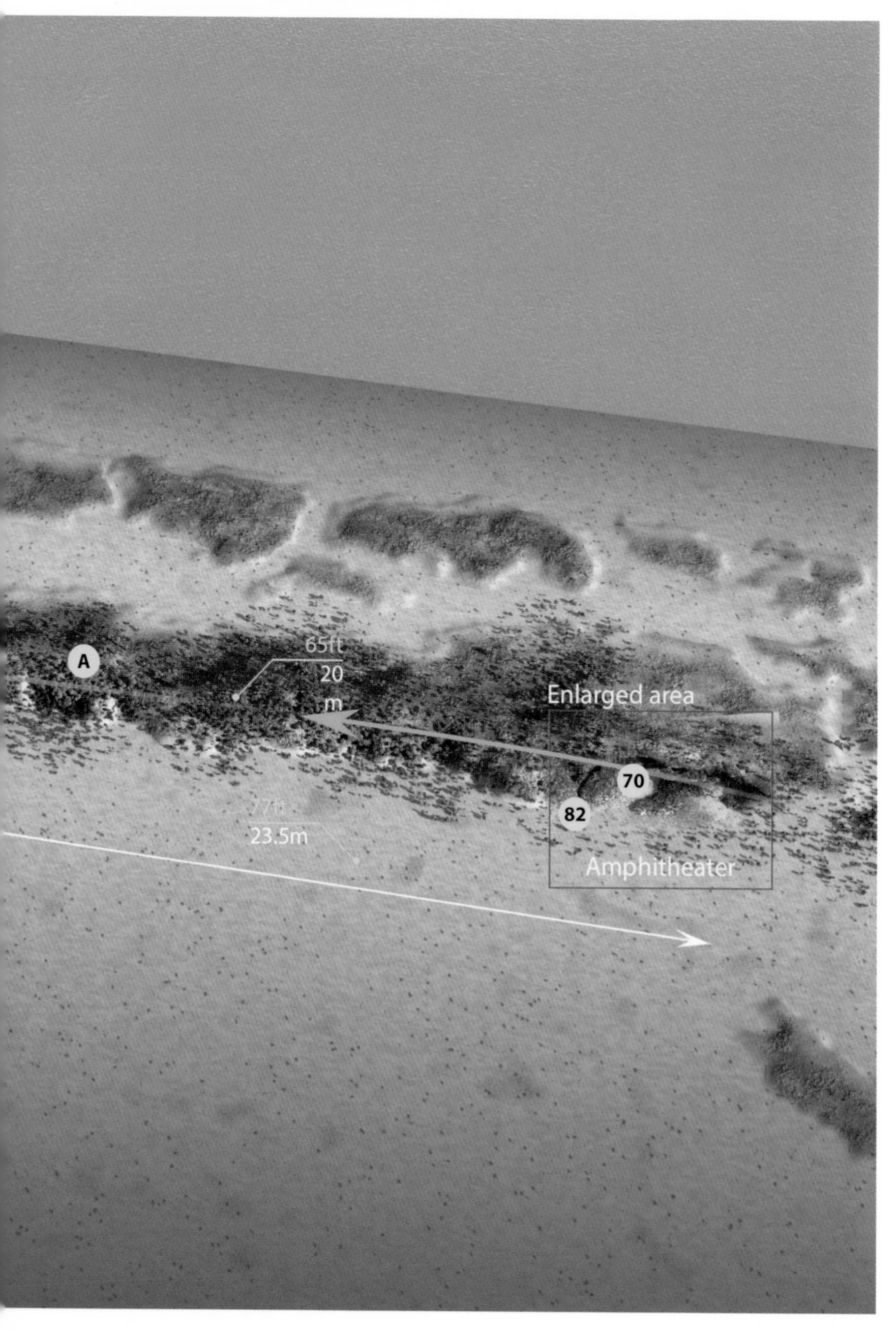

dive operator. After exploring the initial section of ledge and its swim-through, divers often drift north while inspecting the undercut areas of the wall for marine life. Do not forget to keep an eye on the sand for the large southern stingrays that frequent this dive site.

About 25 minutes into the dive with a moderate current, divers may reach a large undercut area that shelters goliath grouper. After this spot, the reef flattens out in some sections, becoming much less wall-like. The final section of this reef is the "donut hole," which can usually only be reached by nitrox

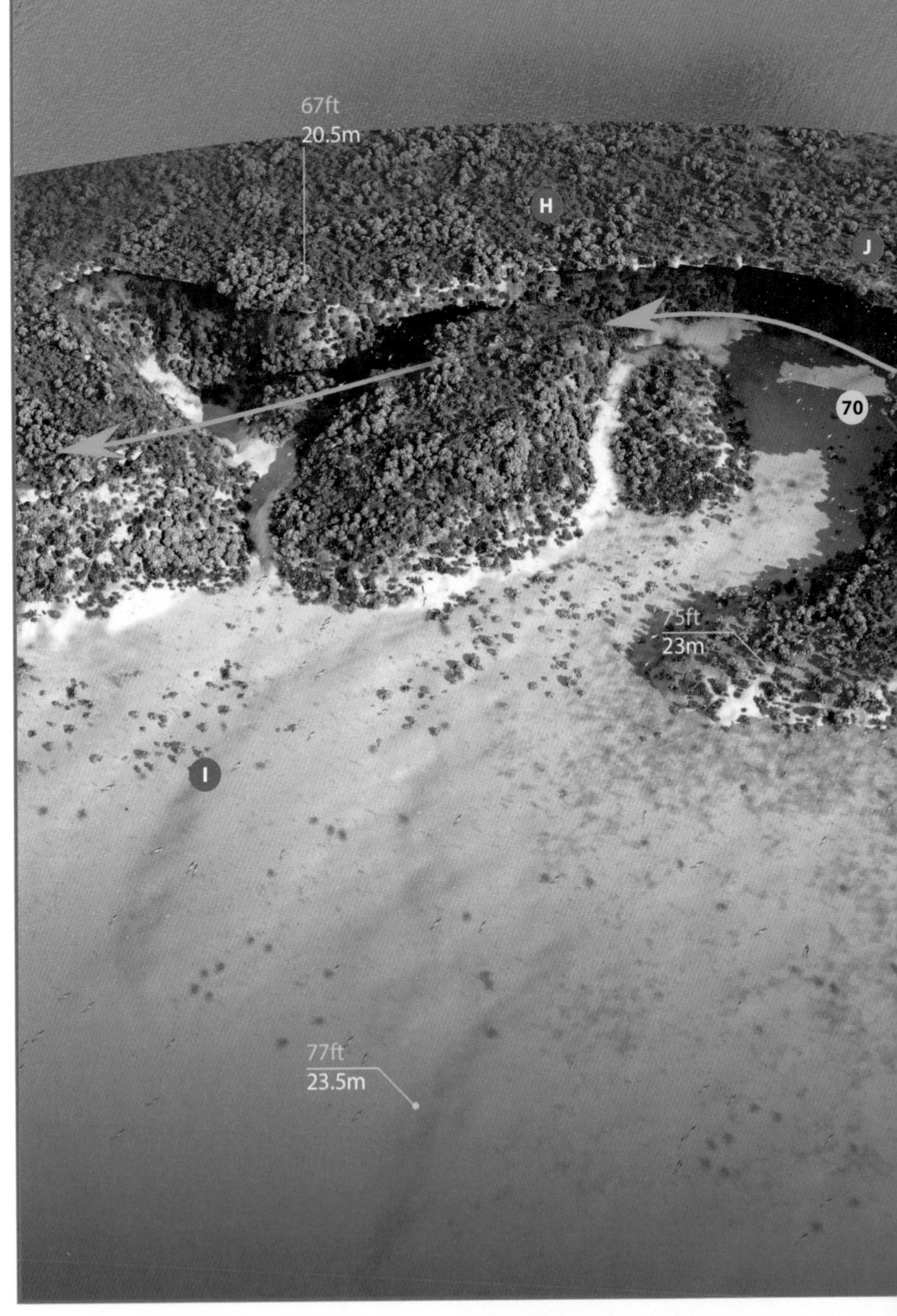

divers when the current is strong enough. Divers using air are unlikely to get this far.

Enlarged area: The amphitheater near the start of the dive includes a swim-through that often shelters goliath grouper.

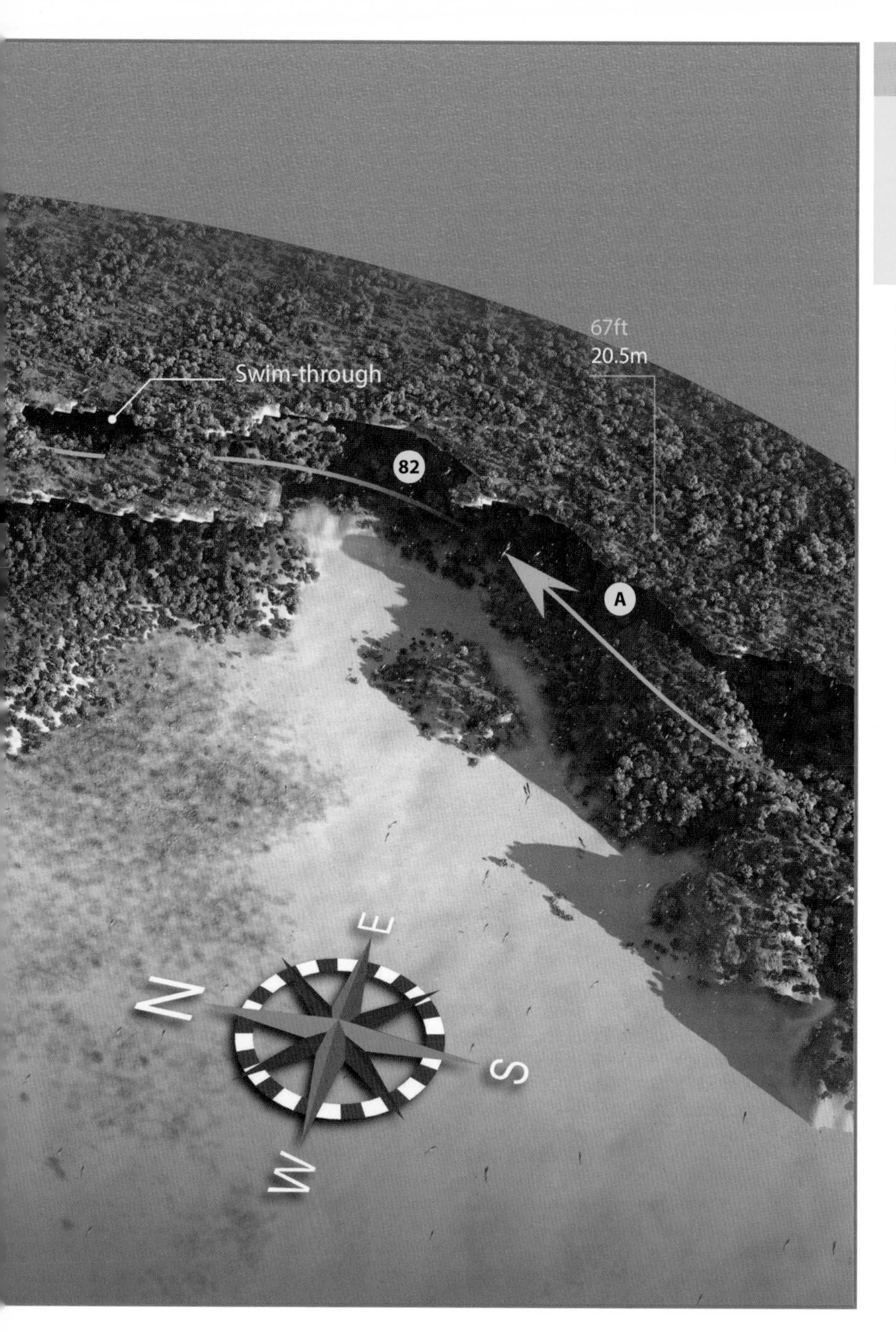

Other species commonly found at this site: N 1 12 31 32 34 41 52 57 67 77 84 86 87 89 90

Scarface

Difficulty ● ● ○
Current ● ● ●
Depth ● ● ○
Reef ★☆☆
Fauna ★★☆

Access about 20 mins from Jupiter Inlet

Level Open Water

Location

Juno Beach, Palm Beach County
GPS: 26°54'51.7"N, 80°01'00.1"W

Getting there

Scarface is a section of reef ledge that runs roughly parallel to the coastline just 2.8 miles (4.5 kilometers) off shore from Jupiter. Scarface sits 3.5 miles (5.7 kilometers) southeast of the Jupiter Inlet, and is only accessible by boat due to its distance from shore. The best way to reach the site is through one of the local dive operators that uses Jupiter Inlet.

Access

Scarface is accessible to divers of all experience levels. Most divers explore this site as a drift dive. Visibility is generally good, but currents can be strong. The currents can help divers explore more of the ledge on their dive, but it also means this site may be more suited to experienced divers.

The reef top is interesting to explore, but the complex network of crevices and channels that cut between the boulders along with the ledge undercuts closer to the sand, are often the most rewarding for divers. However, close buoyancy control under drift diving conditions is an important skill if your planned profile takes you to the deeper part of the site. There are no mooring buoys anchored to this site and divers should have a surface marker buoy with them when they dive.

> **SAFETY TIP**
>
> Extend your bottom time by using nitrox if you plan to explore the complex reef habitat in the deeper part of this site. Divers using air can extend their dive by staying above the ledge, at a depth of 65 feet (20 meters).

Scarface was named after a green moray eel.

Description

Local dive operators named this section of ledge Scarface after a resident green moray eel with a large scar on its face. The site consists of a long stretch of reef ledge, approximately 1 mile (1.6 kilometers) in length. It is the northern section of a contiguous stretch of reef that spans more than three miles. This section of the ledge bottoms out on a sand and rubble seafloor at a depth between 80 to 85 feet (24 to 26 meters). It is slightly shallower than Captain Mike's, located over a mile to the south. Another site, Area 51, sits directly between Captain Mike's and Scarface, offering similar experiences.

The ledge at Scarface has a relatively flat top along most of its length, which varies in depth between 65 and 70 feet (20 and 21 meters). The reef top is sparsely covered with a mix of macroalgae, soft and hard corals, and sponges – none of which manage to grow very big due to the strong currents typical along this stretch of coastline. Shallow depressions along the reef top shelter the occasional school of snapper but the habitat is not particularly complex on top. As a result, the main attractions of this drift dive are the many undercuts along the reef ledge. These undercuts range from just a foot (0.3 meters) to as much as 6 feet (2 meters) in some places. Divers will typically see schools of nocturnal squirrelfish and grunts sheltering under the ledge.

The ledge near the south of this site is more pronounced than in the north, where the reef flattens out and the ledge all but disappears in a few spots. To the south, the edge of the reef is littered with large boulders that have broken off the ledge, providing complex habitat that supports barracuda, nurse sharks and goliath grouper.

Wet Lizard Photography/Shutterstock ©

The height of the ledge varies from just a few feet to as much as 15 feet (4.5 meters) at one point, where the ledge forms a tall cliff. Divers reach this cliff at around the 15-minute mark in their dive, depending on the strength of the current and their starting point. The cliff tops out at a depth of 65 feet (20 meters) with a vertical face that disappears into the sand at a depth of just over 80 feet (25 meters).

The boulders and protected face of the ledge are home to a variety of marine life, including sea fans, sea whips and hard and soft corals. Divers frequently spot many species of angelfish, including French and Queen, patrolling the ledge. Divers can also see snapper, honeycomb cowfish and pretty much every species of grunt that can be found in Florida waters. And though the namesake of this site has long since passed, there are plenty of green morays in the jumble of crevices along the ledge. Hawksbill sea turtles can be seen munching on sponges while

loggerheads hunt for crustaceans near the edge of the ledge.

Route

Most divers follow a route that drifts from south to north along the ledge with the prevailing currents. They often take time to inspect the edges and undercuts if current permits. The reef line tends to take on a scalloped shape, with the leading edge at times pushing out toward the west (toward shore) and at times receding back toward the east (out to sea). Divers will know they are reaching the end of the dive when the reef

Other species commonly found at this site: 3 11 13 15 17 19 22 31 42 48 62 64 65 67 68 85

begins to flatten out. The ledge curves around to head in a northwest direction near the end of the dive and leads to a ledge known as Captain Kirle's.

Enlarged area: Nearly a third of the way through the drift dive, divers will encounter a tall promontory. The reef sports a series of deep undercuts in the ledge leading up to the promontory.

75ft
23m
38
29
80ft
24.5m
C
S
E
W
N

Area 51

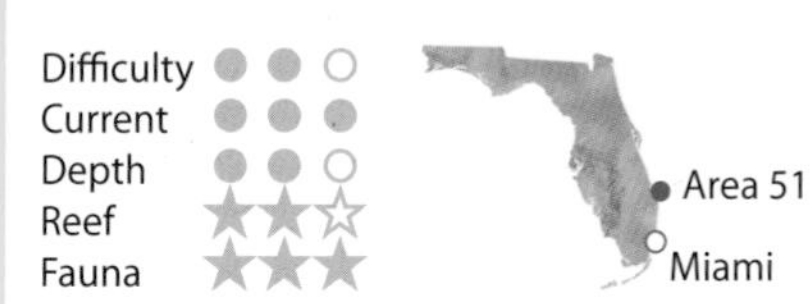

Access about 38 mins from Jupiter Inlet

Level Open Water

Location

Juno Beach, Palm Beach County
GPS: 26°52′50.0″N, 80°00′53.6″W

Getting there

Area 51 is a section of reef ledge that runs parallel to shore just 2 miles (3.5 kilometers) off Juno Beach. It sits just under 6 miles (10 kilometers) south of the Jupiter Inlet, and is only accessible by boat due to its distance from shore. The best way to reach Area 51 is through one of the local dive operators that use the Jupiter Inlet.

Access

This site is accessible to divers of all experience levels. However, the depth means that divers will have limited bottom time to fully explore the ledge. Area 51 is therefore more accessible to divers certified to dive with nitrox. Most divers explore this site as a drift dive. Visibility is generally good and currents can be strong, which helps divers explore more of the ledge on their dive. There are no mooring buoys anchored to the reef and divers should have a surface marker buoy with them.

Description

There are no aliens at Area 51, nor are there any downed alien spacecraft. According to some sources, the site earned its name as a play on the more famous Nevada site of the same name. Others hold that its name originates from the 51 minutes it took a local boat to arrive at the site. Either way, this drift dive offers plenty of ledge for divers of all experience levels to explore. It is located 0.75 miles (1.2 kilometers) north of the dive site known as Captain Mike's (described on page 82) and represents a continuation of the same reef system.

The ledge at this site ranges in height from 10 to 15 feet (3 to 5 meters) – slightly taller than it is at Captain Mike's. The reef crests at a depth of between 70 and 75 feet (21 to 23 meters) and bottoms out at sand at a depth of 90 feet (27.5 meters). The ledge has fractured along much of its length here, with large sections having split off the main reef. The resulting cracks and crevices offer plenty of shelter for reef fishes large and small, ranging from solitary goliath grouper to schools of grey snapper.

Large schools of creole wrasse can be found at Area 51.

Divers will likely see Atlantic spadefish swimming above the reef while barracuda often shelter down below the lip of the ledge, swimming in place in the current. Meanwhile, green moray eels, Spanish hogfish, bar jacks, creole wrasse,

RELAX & RECHARGE

Guanabanas is an open-air, island-themed restaurant nestled on the Intracoastal across from Burt Reynolds Park in Jupiter. Originally opened in 2004 by surfers, the sandwich shop has grown into a local institution with a fantastic mix of food, drinks and live music. It boasts extensive breakfast, lunch and dinner menus that feature salads, tacos, steaks and an incredible range of fresh seafood, including macadamia and coconut encrusted catch of the day. Guanabanas goes to great lengths to ensure its menu items are fresh, locally sourced and sustainable. Even their wine list is sustainable, including only organic certified wines from carbon neutral vineyards. Visit: **Guanabanas.com**

Polly Dawson/Shutterstock ©

blue parrotfish and French angelfish help round out the broad diversity of species at Area 51. To top it off, loggerhead turtles and reef sharks regularly cruise through the area, so divers should keep their eyes open for these visitors as well.

The reef itself is slightly less interesting than the marine life it supports. The top of the reef is dominated by macroalgae – where the strong currents may prove too strong for most corals and sponges to grow to any meaningful size. But the complex habitat of the ledge supports a heavy coverage of smaller sponges and corals, including black corals. Gorgonians and soft corals have colonized the sheltered areas of the ledge, and most surfaces are densely colonized by some form of colorful reef life.

Route

Most dives begin at the southern end of this stretch of reef, which allows the current to carry divers to the north, following the ledge. The exact starting point does not really matter since

the ledge is easy to follow and provides plenty to see along its entire stretch. The dive ends whenever divers reach their maximum bottom time or air reserve.

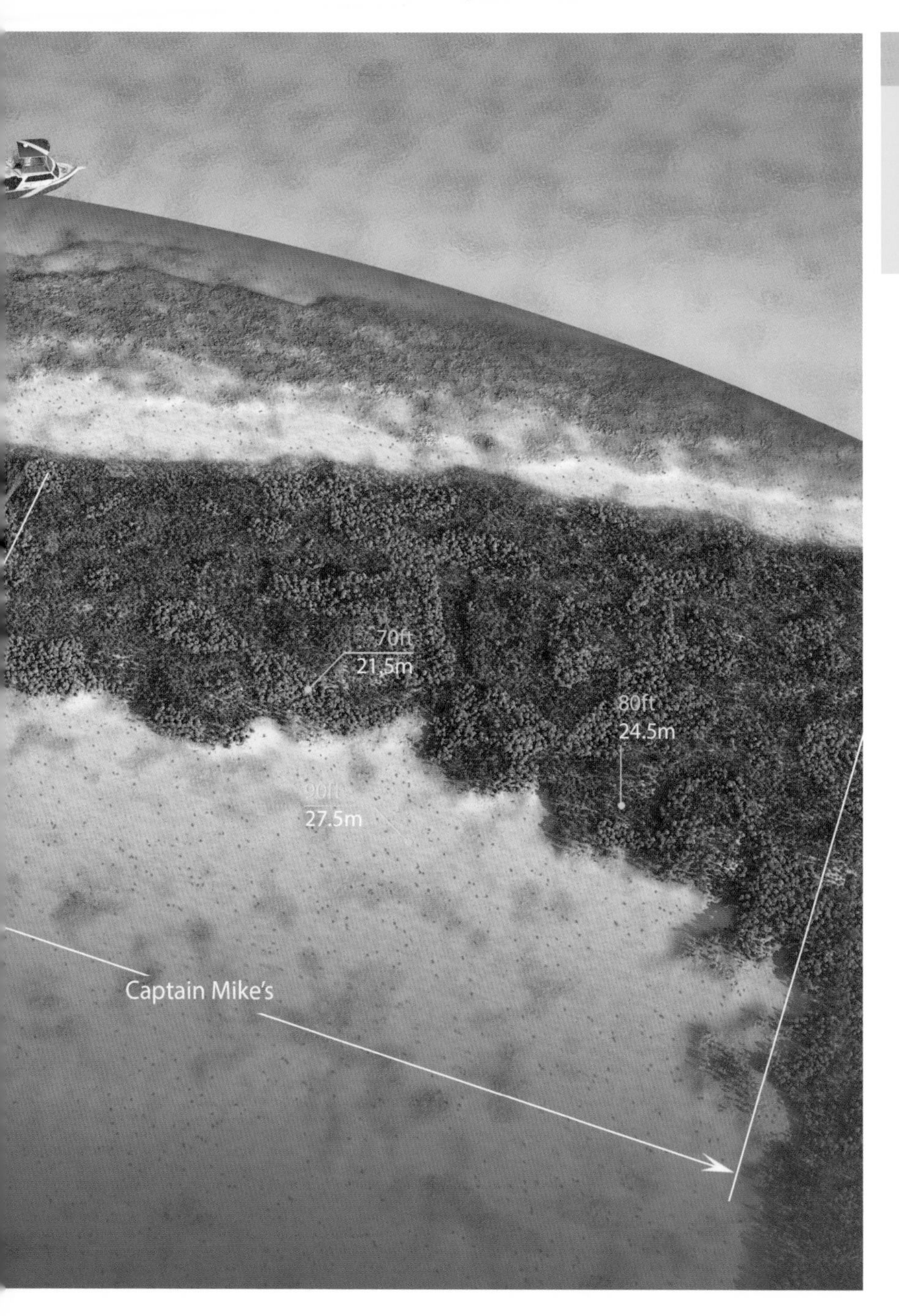

Other species commonly found at this site: **1** **2** **4** **11** **21** **22** **30** **31** **33** **36** **40** **43** **48** **52** **76** **89**

Enlarged area: A typical section of reef ledge that occurs along Area 51. The ledge is quite tall in places, and the ledge itself features a number of cracks and crevices that shelter plenty of lobsters and other species, with loggerhead turtles resting on top of the ledge.

79ft
24m
74
87ft
26.5m
27
12
70
45
30
E
N
S
W

Captain Mike's

Difficulty ●●○
Current ●●●
Depth ●●○
Reef ★☆☆
Fauna ★★★

Access about 40 mins from Jupiter Inlet

Level Open Water

Location

Juno Beach, Palm Beach County
GPS: 26°52'12.3"N, 80°00'49.5"W

Getting there

Captain Mike's is a section of the reef ledge that runs parallel to shore just 2 miles (3.5 kilometers) off Juno Beach. It sits 6 miles (10 kilometers) south of the Jupiter Inlet, and is only accessible by boat due to its distance from shore. The best way to reach Captain Mike's is through one of the local dive operators that use the Jupiter Inlet.

Access

This site is accessible to divers of all experience levels. However, the depth means that divers on air will have limited bottom time to fully explore the ledge. Captain Mike's is therefore more accessible to divers certified to dive on nitrox. Most divers explore this site as a drift dive. Visibility is generally good and currents can be strong, which can help divers explore more of the ledge on their dive. There are no mooring buoys anchored to the reef and a surface marker buoy should be used.

Description

Captain Mike's is a dive site off the coast of Juno Beach named after Captain Mike Hoffman, a long-time local captain who passed away in 2019. The site was added to the list of local dive sites in 2006, shortly after detailed data on the topography of coastal Florida waters entered the public domain. Armed with this new data, dive operators began to search for new sites with the potential to offer divers excellent bottom complexity at accessible depths. Captain Mike's does not disappoint.

The site consists of over 0.6 miles (1 kilometer) of reef ledge that measures between 6 to 10 feet (2 to 3 meters) in height. The ledge is undercut along most of its length, with large boulders and collapsed sections that provide an incredible array of complex habitat that has attracted many species of reef fish and other marine organisms. Massive schools of grey snapper and numerous species of grunt cluster among the boulders, while Queen and French angelfish patrol the ledge. Goliath grouper regularly hang out along this stretch of reef, including a resident grouper nicknamed Tug, or Tuggle, depending on who you ask.

Sea turtles are common visitors to local reef ledges.

The focus of this dive is unquestionably the ledge and the various undercuts that dot its length. The reef crests at a depth between 65

and 70 feet (20 and 21 meters), while the sandy seafloor to the west of the ledge bottoms out at 90 feet (27.5 meters). The top of the reef offers little of interest to most divers, as the surface is sparsely covered by a mix of turf and macroalgae, gorgonians and a few soft corals. Sponges, predominately boring and tube sponges, dot the surface as well. The strong and steady currents at this site likely keep the surface of the reef crest relatively clean of meaningful hard coral growth.

By comparison, the ledge offers a bounty of marine life for divers to observe. Sea fans, gorgonians and encrusting corals abound. Large barracuda shelter in the protection of the ledge, along with numerous cleaning stations operated by juvenile Spanish hogfish. Sergeant majors, wrasses and parrotfish, including midnight, blue and rainbow, are seemingly everywhere along the ledge, as are butterflyfish and the occasional smooth trunkfish. This site is also known to host

Fine Art Photos/Shutterstock ©

its share of turtles, including loggerheads, greens and hawksbills.

Route

Most dives begin on the south end of this stretch of reef, which allows the current to carry divers along the ledge toward the north.

Most operators drop divers into an aquarium-like bowl that measures nearly 100 feet (30 meters) across, where a section of ledge juts out toward the west (toward shore). From there, divers follow the ledge to the north, taking time to inspect the edges and undercuts for sheltering squirrelfish and moray eels if the current permits.

About 10 minutes into the dive, depending on the strength of the currents and the exact starting point, divers will encounter a flattened section of the reef with a sandy top and a low ledge with a single undercut. Shortly after this stretch, the ledge rises and extends out to the west, with a large undercut and space between the ledge and the boulder. This location is where you are likely to find Captain Mike's friendly, resident, goliath grouper.

About 5 to 10 minutes after the grouper cave, the reef ledge once again flattens out, with fewer species present. After a few more minutes of drifting with the currents, the ledge and complexity returns, as does the variety and diversity of reef life. Continuing to the north with the current, divers may encounter sea turtles along with reef and nurse sharks as they eventually cross over into a stretch of reef named Area 51, which marks the end of the dive.

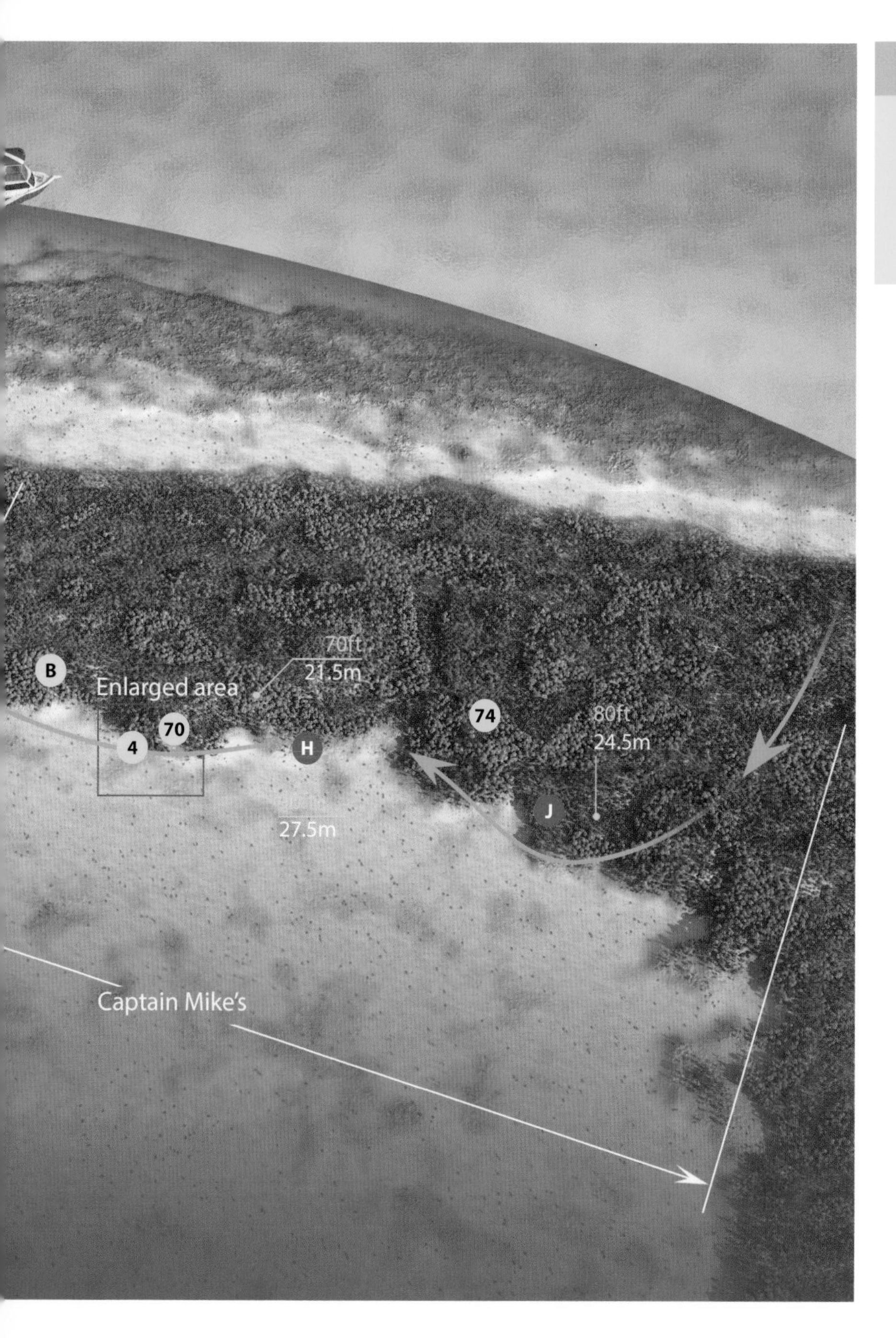

Other species commonly found at this site: 3 4 5 11 17 18 21 27 55 58 63 74 76 79 81 84

Enlarged area: Just past the midway point in the dive, the height of the ledge increases to form a cliff. As the ledge rounds a bend it separates into two distinct overhangs. Goliath grouper often take shelter near here.

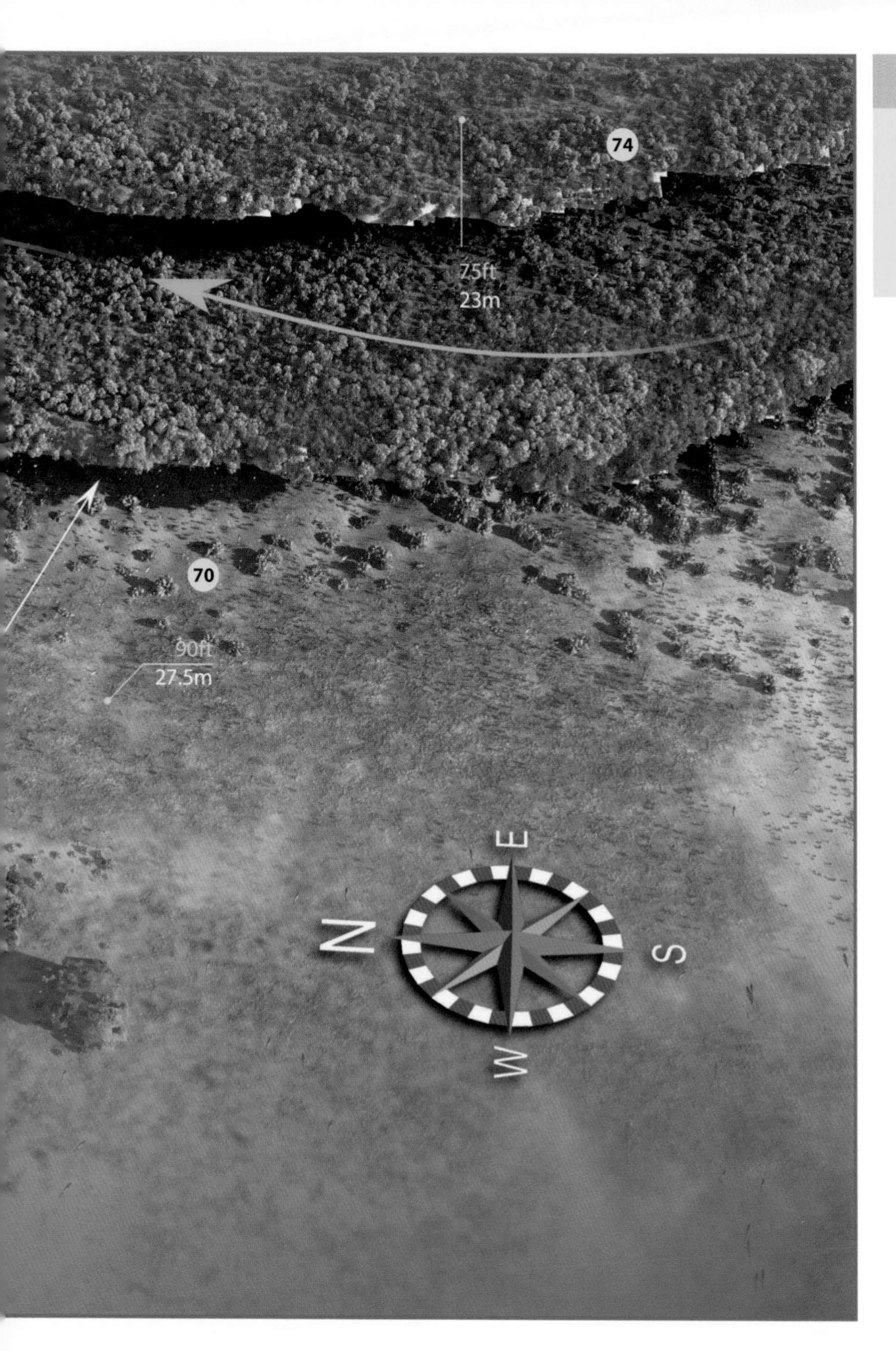

Other species commonly found at this site: 3 7 10 12 14 22 30 32 33 34 37 39 62 71 75 91

Shark Canyon

Difficulty ●●○
Current ●●●
Depth ●●○
Reef ★★☆
Fauna ★★☆

Access about 39 mins from Lake Worth Inlet

Level Advanced Open Water

Location
Juno Beach, Palm Beach County
GPS: 26°51'28.4"N, 80°01'04.7"W

Getting there
Shark Canyon is the name given to two shallow bowl-like amphitheaters cut into the reef ledge that sits 1.7 miles (2.7 kilometers) off the coast of Juno Beach. Located 5.9 miles (9.5 kilometers) north of the Lake Worth Inlet, this site is only accessible by boat due to its distance from shore. The best way to reach it is through one of the local dive centers that operate out of the Lake Worth Inlet even though it is only a slightly longer trip from the Jupiter Inlet.

Access
This site is accessible to advanced divers because of its relative depth. The site is also known for its strong currents and, as a result, it is always explored as a drift dive. Divers typically start the dive by visiting the bowl-like amphitheaters at the southern end of this site, finding shelter from the current behind the reef ledge. Once there, divers can observe the sharks that are frequently spotted at this site, before drifting along the ledge in a northerly direction.

Shark Canyon is best explored during weak to moderate currents as stronger currents can make it difficult to spend enough time in the bowls to observe the sharks. There are no mooring buoys at this site so divers should bring a surface marker buoy with them. Visibility is generally good.

Description
Shark Canyon is located 0.25 miles (0.4 kilometers) west of the reef ledge known as Juno Ledge. It gets its name from the resident sharks, including reef sharks, lemon sharks and even bull sharks, which are often spotted around the two bowl-like amphitheaters at the southern end of this dive site. This area was once popular for shark chumming, before the practice was made illegal within 3 miles (5 kilometers) of the U.S. coast. Even though the chumming has stopped, sharks are often still present here in higher numbers than at other sites. Divers have only to wait a short time in the protection offered by the amphitheaters for the sharks to come cruising by.

The amphitheaters are not fully enclosed. They back onto a relatively flat reef top at a depth of 73 feet (22 meters) and open to the south where the

Several species of shark can be found at shark canyon.

ledge drops off to a depth of 95 feet (29 meters), featuring a number of interesting overhangs and undercuts. In fact, the ledge is interesting enough that when currents are stronger, many dives skip the amphitheaters altogether and focus entirely on the ledge. The southern amphitheater is smaller than its neighbor to the north and slightly deeper at 80 feet (24.5 meters) – the northern amphitheater bottoms out at 78 feet (24 meters). The northern amphitheater is marked by two tiers divided by a small step, with the eastern end slightly shallower at a depth of 77 feet (23.5 meters).

The bottom of each amphitheater is covered in a mix of sand and coral rubble, with a few small solitary coral heads scattered throughout. The walls of the amphitheaters are similar in coverage to the ledges that are scattered throughout the area, with hardy gorgonians and sponges clinging to the reef substrate.

The main ledge continues to the north. As with other ledges along this stretch of the coastline, divers will see plenty of snapper, parrotfish, grunts and angelfish close to the ledge. Barracuda, turtles and grouper are often spotted as well, while stingrays frequent the sand below the ledge.

Route

The drop point for this site is typically just south of the southern-most amphitheater on a flat stretch of algae-covered reef. This position allows divers to navigate toward the main features of this dive site by keeping east of the ledge, on the flat reef crest. Once at the first amphitheater, divers often find it helpful to choose a position close to the reef that provides shelter from the current, but not close enough that they cause damage to the reef or disturb any marine life. Divers often tuck in behind the lip of the amphitheater or hang out behind one of the larger coral heads in its center.

In most cases, the reef sharks arrive within a few

minutes of the divers. Do not chase the sharks; while they often approach to within just a few feet of a motionless diver, they startle easily if chased. Spearfishers should avoid spearing fish close to the amphitheaters, as the history of chumming at this site has led some sharks to be quite bold; they have even been known to steal the catch of spearfishers.

At the second amphitheater, divers can follow the same strategy of settling close to the bottom and allowing the reef sharks to approach on their own. Moving north beyond the amphitheaters, the dive continues along the main ledge, where grouper, sea turtles and a variety of other marine life can be found.

SHARK CANYON

Other species commonly found at this site: I J 2 3 5 6 7 8 11 15 19 22 31 44 48 68

Danny & Atlantis

Difficulty ●●○
Current ●●●
Depth ●●○
Reef ★★☆
Fauna ★★☆

Access about 11mins from Lake Worth Inlet

Level Open Water

Location

Riviera Beach, Palm Beach County
GPS (*Danny*): 26°47'37.4"N, 80°01'05.4"W

Getting there

This dive site represents a large area, and parts of it can be explored in a single drift dive. In total, the site covers an area that stretches nearly 850 feet (260 meters) from south to north and includes several wrecks, several statues and a large amount of concrete debris, including bridge spans and pilings. It is 1.6 miles (2.6 kilometers) north of the Lake Worth Inlet, or an 11-minute boat ride. It is best reached through one of the area's local dive operators.

Access

There is no permanent mooring buoy on any part of this site, so local operators often put divers in the water slightly up current from the Danny McCauley Memorial Reef with enough time to descend so they do not overshoot the wreck. The dive is done most commonly as a drift dive from south to north with the prevailing currents. Currents can be strong, which can make it challenging to get a close look at the statues and other structures at this site. Visibility is generally good, although an outgoing tide can push murky water out of the inlet and through the site, reducing visibility to mere feet at times. The site is accessible to divers of all skill levels.

Description

This site was created over the course of a decade or more. The county deployed the artificial reefs for Atlantis first in 2000 and 2001, which is also known as the Royal Park Bridge site and then added additional artificial reefs over a decade later. The center of the site houses concrete bridge spans and other debris generated from the demolition of the Royal Park Bridge – hence the alternate name – while two barges and two tugboats are placed around the edges to provide additional opportunities for exploration.

Chief among these artificial reefs is the Danny McCauley Memorial Reef (often referred to simply as the *Danny*), which is a Canadian-built ice tug that sits to the south of Atlantis and slightly to the west. Originally named *Pocahontas*, the 110-foot-long (33.5-meter) tug was built in 1944 during the wartime shipbuilding push. Not much is known about *SS Pocahontas*, which is not unusual for a simple tug, although she reportedly languished in the Miami River for 35 years before being acquired for the specific purpose of sinking her as an artificial on February 22, 2013.

The wreck sits upright on a sandy seafloor at a depth of 78 feet (24 meters) and stands as a memorial to Danny McCauley, an avid diver who died in a car accident at the age of 17, one year to the day before the *Pocahontas* was scuttled. Large letters spelling out DANNY are welded to the stern of the boat and a memorial cross stands on the superstructure, near the bow, to greet divers as they approach from the south. The McCauley family and friends were instrumental in raising the funds necessary to acquire, prepare and sink the wreck as an artificial reef. The wreck is heavily encrusted with soft corals and algae. Limited penetration is possible, and holes were cut in the hull to provide easy access.

Nearly 120 feet (36.5 meters) to the east of the *Danny*'s stern – off the port side – sits Andrea's Reef, a collection of small statues that originally included a sea turtle, stingray (although this is no longer visible), four pairs of short Roman columns and a 60,000-pound, life-sized mermaid statue featuring the likeness of Andrea Torrente, a three-time breast cancer survivor after whom the artificial reef is named. The columns form a square positioned roughly 15 feet (5 meters) away from the central mermaid statue. The whole assortment of statues has become heavily encrusted in macroalgae since it was deployed on September 15, 2016.

Stretching to the north from Andrea's Reef is a collection of artificial reef material. The material starts out as scattered boulders, small in size and spread apart, but the debris gradually gets larger and closer together toward the north. This is the beginning of what is referred to as Atlantis, a large area of complex habitat ideal for both large and small marine life. The northern section of the debris field includes long concrete sections of what was once the Royal Park Bridge. These bridge sections increase in density until the end of the debris field, where they are stacked on top of one another like a giant's game of pick-up-sticks. This area is also marked by large concrete box-like structures that were part of the bridge. One such massive box, measures 18 feet by 12 feet (5.4 meters by 3.6 meters) and is a key landmark along the route to the *Spud Barge*.

The concrete structures are heavily encrusted in soft and hard corals as well as sponges. Numerous juveniles of a variety of reef fish species can be seen swimming through the complex habitat,

Peter Leahy/Shutterstock ©

Spotted scorpionfish are often found in the bridge debris of Atlantis.

including tomtates and other grunts, sergeant majors, surgeonfish, snapper and wrasses. It is also not uncommon to spot a solitary goliath grouper hanging out in the current.

The *Spud Barge* is a 120-foot-long (36.5-meter) flat-bottomed barge with holes in the sides of its hull that expose its interior cross braces. The inside of the hull allows for penetration; divers can pass through the entire hull from one side to the other in full sight of the exit points, but the space is a little tight and can make some divers feel claustrophobic.

This is a favorite location for breeding aggregations of goliath grouper during the summer months and the wreck teams with porkfish, grunts and butterflyfish. The hull is heavily encrusted with hard and soft corals and a few errant sponges. A large granite block sits on the hull near the southern end of the barge while other blocks are strewn about the nearby sand.

To the west of the overturned *Spud Barge*, and due north of the *Danny*, lies the wreckage of another, smaller barge and a small tug. The overturned hull of the 65-foot-long (20-meter)

barge rests largely on top of the bow of the 50-foot (15-meter) tug. Both are heavily encrusted in macroalgae interspersed with sponges and hard and soft corals. Porkfish and sergeant majors are plentiful on this wreck, along with a variety of snapper and grunts, and even goliath grouper. The stacked nature of the two wrecks means there is a small space under the barge that divers can peer into with the help of a flashlight. But given the unstable conditions of the two wrecks, full penetration is not recommended.

Other species commonly found at this site: L 3 4 7 11 16 21 23 26 30 37 40 43 47 48

Route

It is virtually impossible to visit all the elements of this large site in a single dive. The positioning of the small barge and tug relative to the *Spud Barge* means that with any kind of current it would require extensive effort to cut back across to see both barges, not to mention the mermaid statue and Atlantis debris. Rather than let this discourage divers, however, it provides all the more reason to visit this site on multiple dives.

The most common route involves divers exploring the *Danny* before cutting across the current in a diagonal path to visit Andrea's Reef.

Name:	Danny McCauley Memorial Reef	**Construction:**	Canada 1944
Type:	Tug	**Last owner:**	Unknown
Previous names:	*SS Pocahontas*	**Sunk:**	February 22, 2013
Length:	110ft (33.5m)		
Tonnage:	Unknown		

From there they typically drift with the current through the Atlantis debris field toward the *Spud Barge*. It is nearly impossible to navigate the debris field based on individual boulders and bridge pieces, although a few key landmarks can help orient divers. Two main landmarks include the large concrete box and the collection of bridge pieces pointing toward the surface at random angles. The latter earned a nickname of the Fortress of Solitude from local operators due to its similarity to Superman's Arctic fortress of the same name. Most divers begin their ascent to the surface after the *Spud Barge*.

Danny position

Alternatively, divers can head north after exploring the *Danny*, directly off the stern. After crossing some scattered debris, they will encounter a large funnel-like object that often shelters a goliath grouper. From there, divers continue drifting with

the current until they encounter the small barge and partially crushed tug. After exploring the double wreck, most divers then ascend to the surface.

Danny's funnel
J
75ft
23m
Danny's funnel position
Spud barge
N
W
E
S
Royal Park Bridge
Tug and Small barge
Andrea's Reef
Funnel
Danny

S
W
E
N
65ft / 20m
65ft
20m
Bow
70
5ft / 1.5m
Small Barge position
Spud barge
N
W
E
S
Royal Park Bridge
Tug and Small barge
Andrea's Reef
Funnel
Danny

Name:	Tug and Small barge	**Construction:**	Unknown
Type:	Tug and Barge	**Last owners:**	Unknown
Previous names:	n/a	**Sunk:**	2000/2001
Lengths:	Tug: 50ft (15m) Barge: 65ft (20m)		
Tonnage:	Unknown		

Danny
Bow
76ft
23m
38
70
120ft / 36.5m
76ft
23m
Spud Barge position
Spud barge
N
W
E
S
Royal Park Bridge
Tug and Small barge
Andrea's Reef
Funnel
Danny

Name:	*Spud barge*	**Last owner:**	Unknown
Type:	Barge	**Sunk:**	2000/2001
Previous names:	n/a		
Length:	120t (36.5m)		
Tonnage:	Unknown		
Construction:	Unknown		

Ocean Reef Park

Difficulty ●○○
Current ●○○
Depth ●○○
Reef ★☆☆
Fauna ★☆☆

Access about 15 mins from West Palm Beach
about 1 min from shore

Level n/a

Location

Singer Island, Palm Beach County
GPS: 26°47′37.3″N, 80°01′52.4″W

Getting there

Ocean Reef Park is a shore-accessible stretch of fringing reef that runs along the coast of Singer Island – a barrier island just north of Riviera Beach. The reef is located along the stretch of beach associated with the county park of the same name, Ocean Reef Park. There is ample parking associated with the park, along with bathroom and shower facilities. To get there, cross over the Blue Heron Bridge from US Hwy 1 in Riviera Beach. Once on Singer Island, follow Hwy A1A as it curves left and heads north. Drive another 0.7 miles (1.1 kilometers) and the park will be on the right, immediately after the Palm Beach Marriott at Singer Island Beach. The park's address is 3860 North Ocean Drive Riviera Beach, 33404.

Access

Ocean Reef Park is accessible to snorkelers of all experience levels. The beach is a short walk from the parking lot over the wooden steps and boardwalk that cut through the lush shoreline vegetation. The beach has a lifeguard on duty during normal hours. Snorkelers can check in at the lifeguard tower to get a sense of the conditions before they get in the water. The shallowness of the reef makes it best to access this site during high tide and when the breaking waves are smaller. Snorkelers should enter the water where there is sand, holding their fins in their hands. It is important to pay attention to foot placement to avoid stepping on the reef or any organisms, which could result in injury to the snorkeler or damage to the reef. Once snorkelers have made their way to waist-deep water, they can put on their fins and then assume a natural snorkeling position – floating on the surface of the water until they get their bearings. A dive flag is not necessary at this site provided you stay within the guarded area marked by buoys.

Description

The Ocean Park Reef includes patches of reef and small ledges that dot a stretch of coastline measuring close to 1,000 feet (305 meters) in length. The area is bracketed to the north by an outcropping of limestone and to the south by a flare of shallow fringing reef, which sits directly in front of the Marriott hotel. Snorkelers can explore the reef all the way into the surf zone – being careful not to let the waves push them into the sharp rocks of the reef – and all the way out to the buoys that float just off shore, warning boats and jet skis away from the swimming area.

There are three main sections of reef at this site, the shallow flame-shaped fringing reef to the south, a fan-shaped section of reef in the middle

A snorkeler checks out a shallow natural reef.

of the area, located between the two lifeguard towers, and a section of small patch reef and rubble that marks the northern edge of the site.

The high wave action and shallow depths mean that only the hardiest of reef organisms can survive here, such as the encrusting fire corals that thrive in these conditions. While not true corals, these organisms sting on contact, which can produce a painful, burning, red welt. While the reef itself may be mostly rocky limestone, it provides ample habitat for juvenile reef species. Snorkelers may encounter numerous species from grey snapper to spotted drums and banded butterflyfish at this site. Chub are common above the reef, while sharp-eyed snorkelers may pick out the cryptic scorpionfish hiding in the sandy rubble. Small dark damselfish frequent this type of high energy environment, while stingrays can often be seeing patrolling the sandy spaces in the deeper water.

RELAX & RECHARGE

Back across the Blue Heron Bridge and a couple of miles (3.2 kilometers) to the north on Hwy 1, lies a pleasant waterfront restaurant called **Frigate's Waterfront Bar & Grill.** The broad menu supports the restaurant's claim to have "something for everyone." You can order everything from burgers and pasta to salad and seafood, including grouper and hogfish. They even have a sushi menu available from Wednesday to Sunday, and a diverse Sunday Bunch menu, considered one of the best in the county. Frigate's also has some great signature drinks, including their version of Dr. James Munyon's Paw-Paw Elixir. Munyon was a typical 19th century "quack" – creator of homeopathic medicines – who was found guilty of fraud several times. He built and operated a hotel on Nuctsachoo (Pelican Island), just to the north of the restaurant – now known as Munyon Island and part of John D. MacArthur Beach State Park. The Paw-Paw Elixir, which was made of fermented papaya juice among other ingredients, was Munyon's most famous concoction, and brought many northerners flocking to his hotel during the early 1900s. Visit: **Frigatesnpb.com**

Andrey Armyagov/Shutterstock ©

Route

A typical route may involve entering the water in the sandy stretch between the southern and middle sections of reef. Snorkelers can then head south to explore the ledge that rings the fringing reef to the south. After exploring that section, snorkelers can make their way to the north pausing to visit the fan-shaped reef and then moving on to the patch reef to the north. Snorkelers might want to consider circling back through deeper waters for the chance to see larger fish, including stingrays that are commonly seen patrolling or resting in the open sand.

Other species commonly found at this site: 0 1 6 9 10 15 16 19 22 24 37 42 46 58 73 78

Ana Cecilia & Mizpah Corridor

Difficulty ●●○
Current ●●●
Depth ●●○
Reef ★☆☆
Fauna ★★☆

Mizpah Corridor
Miami

Access about 8 mins from Lake Worth Inlet

Level Advanced Open Water

Location

Riviera Beach, Palm Beach County
GPS (*Ana Cecilia*): 26°47′05.2″N, 80°00′56.9″W
GPS (*Mizpah*): 26°47′10.8″N, 80°00′57.6″W

Getting there

These two dive sites are almost always explored as a single, particularly long, wreck trek. The Mizpah Corridor was deployed first, and the *Ana Cecilia* was sunk more recently, at the head of the trail. The new addition extends the popular trek to the south. If you consider both sites together, they create a trek that stretches over a third of a mile (over a half kilometer) from south to north and includes a total of five wrecks, two rock piles and one discarded superstructure. The trek is located just an 8-minute boat ride north of the Lake Worth Inlet, and it is best reached through one of the area's local dive operators located near that inlet.

Access

There are no permanent mooring buoys on any of the wrecks in this trek. Local operators will often put divers in the water slightly up current from the trek so they have time to descend to the bottom without overshooting the *Ana Cecilia*. The site is explored most commonly as a drift dive along with the prevailing currents, typically from south to north. Currents can be strong along the entire length of the trek, which is both a benefit and a challenge for divers. Stronger currents mean divers will have a hard time exploring more than one side of any individual wreck, but they also help divers reach all the wrecks in the trek.

Visibility is generally good on this site. However, an outgoing tide will push murky water out of the inlet and straight through the wreck trek, which can reduce visibility. These combined dive sites are best accessed by divers holding Advanced Open Water certification, or equivalent drift and

Andrea Whitaker ©

A diver passes over the top of the *Mizpah*.

DID YOU KNOW?

The Mizpah Corridor trek theoretically continues past the *China Barge* to a debris field known as the Brazilian Docks. From there, divers can continue along to the *Danny* and Atlantis multi-wreck site. The total distance from the *Ana Cecilia* through to the *Spud Barge* (the northern wreck in the *Danny* and Atlantis site) is nearly a mile (1.6 kilometers) and so it is next to impossible for either air or nitrox divers to make it from end to end with any time to spend exploring the individual wrecks. Most divers are satisfied to explore the Mizpah Corridor on one dive and the *Danny* and Atlantis site on another.

wreck diving experience. Diving on nitrox is also recommended, as it is next to impossible to reach the end of the trek within a safe bottom time when diving on air.

Description

Taken together, these two sites, the *Ana Cecilia* and Mizpah Corridor trek, offer divers the opportunity to visit five wrecks – the *Ana Cecilia* in the south, followed by *Mizpah*, *PC-1174*, the *PC-1174* superstructure, the massive hollowed out hull of the *Amaryllis* and finally the *China Barge* (also known as Research Team Reef). The combined trek stretches 0.35 miles (0.56 kilometers) from south to north, and includes two separate piles of rocks, each weighing over 1,260 tons. The two piles were placed to the south and north of the *Amaryllis* in 1994 and are referred to as Habitat Corridors South and Habitat Corridors North, respectively.

The boulders offer divers a convenient navigational aid as they drift with the current, while also providing valuable reef habitat for marine organisms. A mix of gorgonians, sponges and small coral heads blanket each pile, while a range of reef fishes, including sergeant majors, snapper, Spanish hogfish, parrotfish and surgeonfish call these reefs home.

The *Ana Cecilia* (also sometimes spelled *Ana Cecelia*) was originally built in 1972 by U.S.-based shipbuilders Halter Marine. Originally a 170-foot (52-meter) general cargo ship, *Ana Cecilia* had a mixed history. She was reportedly one of the first vessels to deliver humanitarian supplies to the island of Cuba in over 50 years. The supplies were needed after eastern Cuba suffered a direct hit from Hurricane Sandy in 2012. Authorities also suspected she was involved in smuggling drugs between Haiti and the U.S. Extensive surveillance efforts led authorities to set up a sting operation in September 2015, which resulted in the seizure of over 1,100 pounds (507 kilograms) of cocaine – a street value of over $10 million – while she was docked in the Miami River. She is the most recent addition to the trek, sunk as an artificial reef in July 2016, and so supports less coral growth than other reefs on this site. Plenty of reef fish species call the wreck home, however, including porkfish, barracuda, grey snapper and sergeant majors.

The *Mizpah* was a 185-foot (56.5-meter) luxury passenger ship built in 1926 and sunk as an artificial reef in 1968. Widely reported as merely a Greek luxury yacht, she actually has a far more interesting and varied past. She was built in Newport News, Virginia in 1926 for the Navy under the name *Savarona* but was scrapped before she was completed. The Navy sold the incomplete vessel to a wealthy family from Palm Beach, the Cadwalders, who finished her off as a luxury yacht under the name *Sequoia*. The Cadwalders reportedly sold her to James Elverson, who renamed her the *Allegro*.

In 1929, she was sold once again, this time to Commander MacDonald, who outfitted the yacht with state-of-the-art electronics and turned her into a floating home. She primarily plied the waters of the Great Lakes and Florida, but allegedly ventured into Bahamian waters on a few treasure hunting expeditions.

When World War II broke out, the family donated the yacht back to the Navy under the name *PY-29*. The Navy outfitted her with guns and put her to work as an escort vessel, where she hunted U-boats along supply routes in the Atlantic. After the war, the Navy returned *PY-29* to Commander MacDonald, who sold her on to a Honduran shipping company. After nearly two decades of work as a freighter, she ran aground on a reef carrying bananas from Honduras to Tampa, Florida.

Heavily damaged, she was towed to a shipyard, which is where Eugene Kinney, a descendant of Commander MacDonald, found out about her. Eugene purchased the *Mizpah* and ultimately decided to donate her to Palm Beach County's artificial reef program. The *Mizpah*'s original position was chosen to enhance fishing, but the Gulf Stream current reportedly grabbed her as she sank and pushed her into shallower waters – which proved to be a boon for divers.

The next vessel in the corridor is *PC-1174*, a patrol

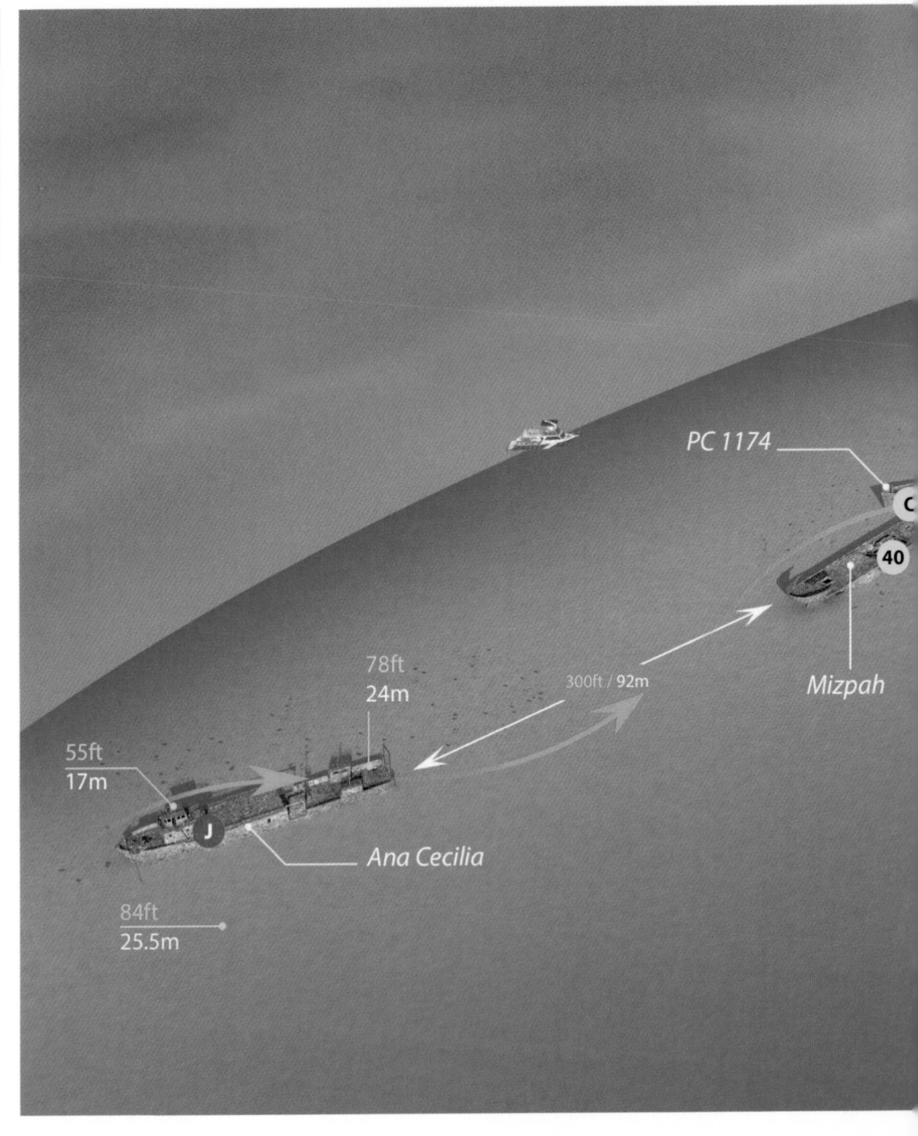

craft scuttled to create an artificial reef the day after the *Mizpah* was sunk. One of a line of Navy sub-chasers built during World War II, *PC-1174* originally served as an escort craft in the Gulf of Mexico and along the southeast coast of the U.S. In June 1944, the Navy refitted her for overseas duty. She spent a year escorting Allied transport ships in the Mediterranean and participated in the invasion of France before returning to the Keys. She was headed to the Pacific Theater when the war with Japan ended.

The Navy put her to use towing bombing and strafing targets for the Atlantic Fleet Air Wing until she was eventually decommissioned in 1947 and mothballed in the St. Johns River in Florida. While there, she was given the name *Fredonia* and sold to private interests in 1957. Members of the Palm Beach Sailfish Club eventually purchased the ex-sub-chaser and donated her to the county's artificial reef program to be sunk, along with the *Mizpah*, in the spring of 1968.

The two wrecks were originally positioned hundreds of yards apart, but over the years, currents from various hurricanes pushed the *Mizpah* northward until she came to rest against the hull of *PC-1174*. Currently the two wrecks form one large artificial reef that is heavily colonized by hard and soft corals, as would

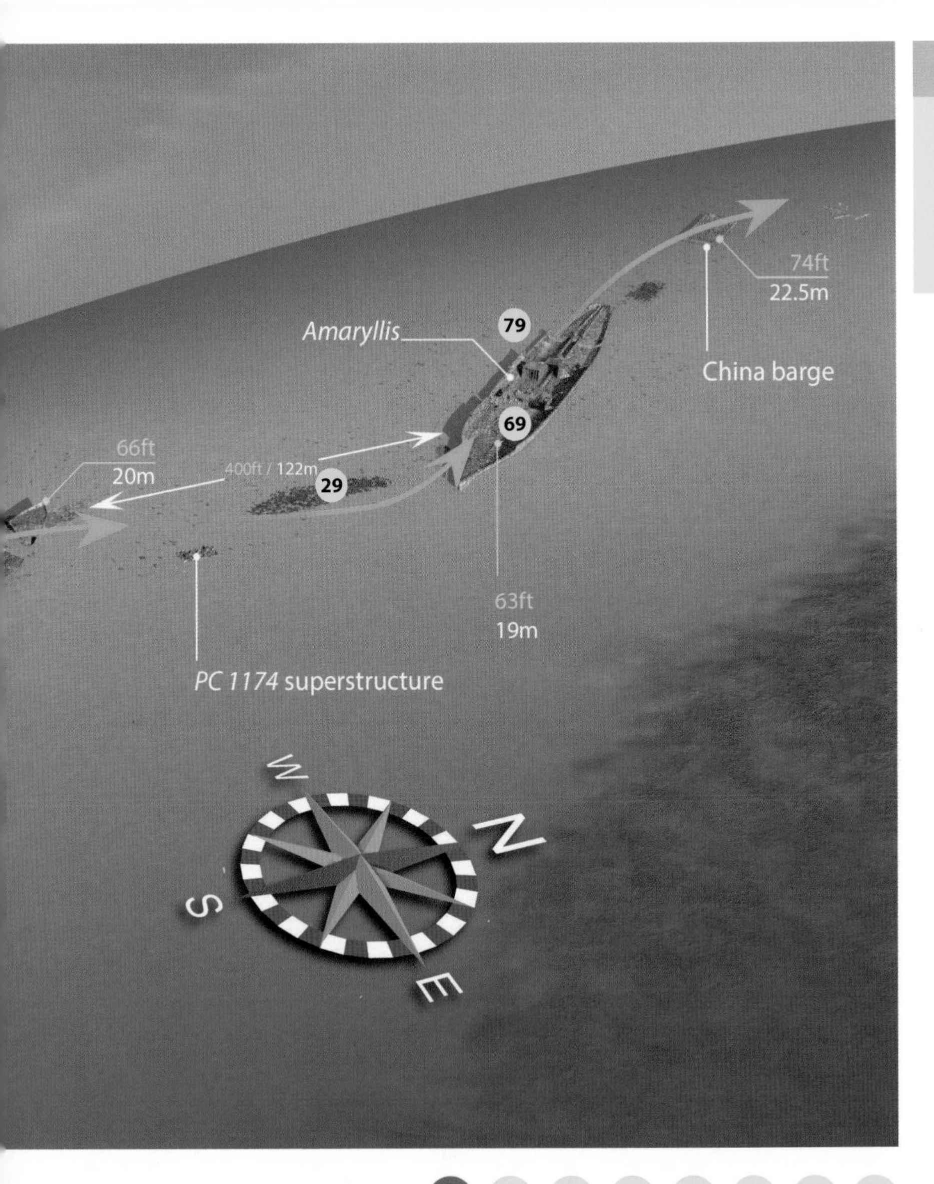

Other species commonly found at this site: N 3 5 6 13 15 21 22 30 31 32 36 37 71 72 91

be expected after a half century underwater. They also support a variety of reef fish species, including Queen angelfish, sergeant majors, blue tangs and goliath grouper. Just to the east of these two wrecks lies the superstructure of *PC-1174*. This small rectangular section of the patrol craft sits on a sandy bottom off to the side of the main corridor. It is not a large structure, but it does offer the chance to see goliath grouper during the summer spawning season.

Beyond *PC-1174* lies one of two rubble piles that bracket the hollowed wreck of the *Amaryllis*. The *Amaryllis* was originally built in Canada as part of a series of Park-class merchant vessels constructed during World War II. Operated by the Canadian government, the Park-class cargo vessels were based on the same designs as Liberty vessels

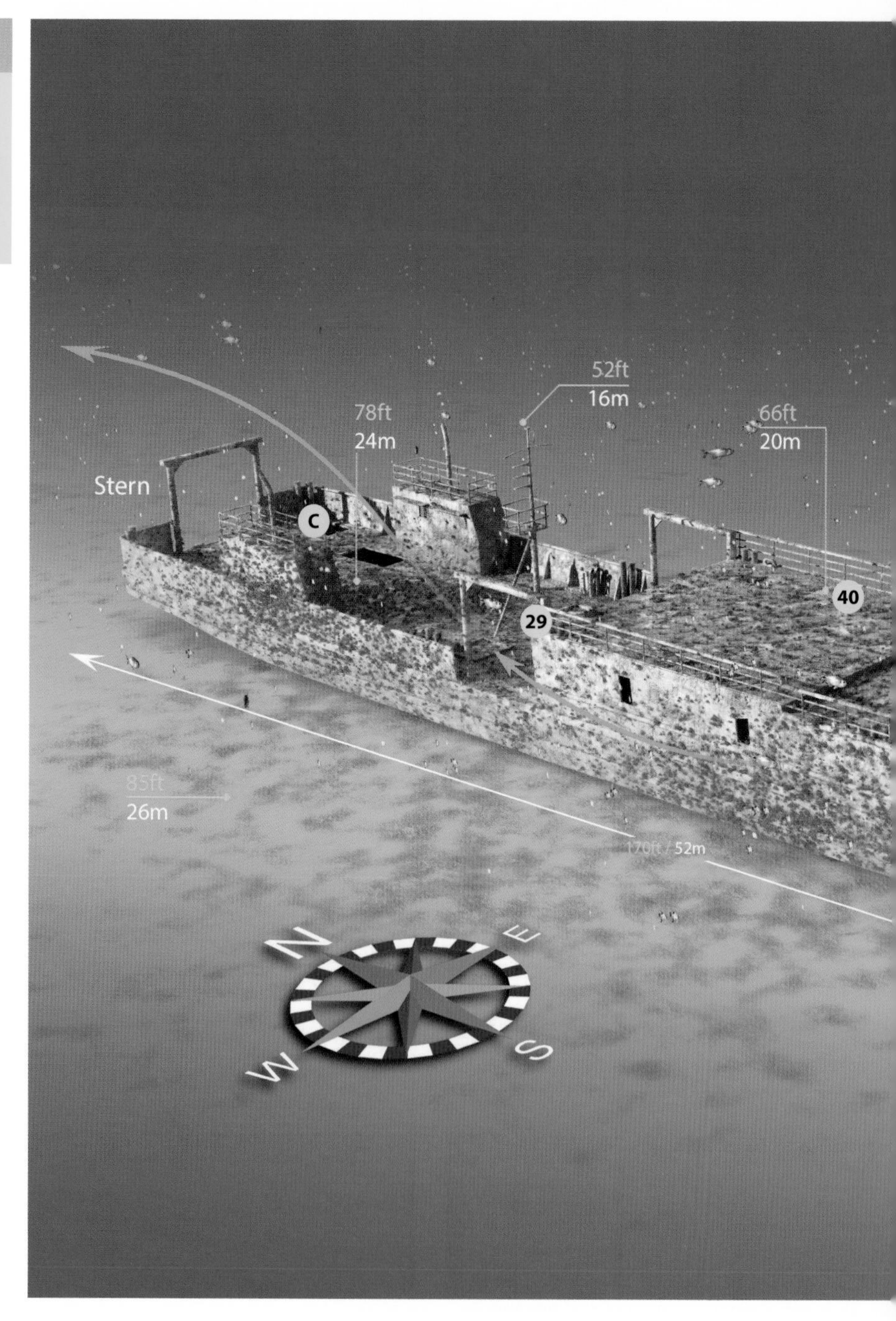

built in the U.S. at the same time. Originally named *SS Cromwell Park*, the new freighter floated out of the Burrard Dry Dock shipyard in Vancouver on Valentine's Day in 1945. After the war, the Canadian government sold her to a private Canadian transport company who renamed her *SS Harmac Vancouver.* Greek shipowners bought her in 1948, renamed her *Amaryllis* and operated her under a Panamanian flag until she ran aground in September 1965.

While en route to Baton Rouge, Louisiana from Manchester, England, *Amaryllis* sought shelter in the Lake Worth inlet as Hurricane Betsy generated

high winds and surf. Steering issues combined with the storm surge prevented her from making safe harbor, however, and she beached just off Singer Island. She reportedly sat on the beach for years, defying all salvage attempts. Eventually deemed enough of a safety concern to warrant the involvement of the Army Corps of Engineers, she was towed out to sea in August 1968 and sunk as an artificial reef, just north of *Mizpah* and *PC-1174*. All that remains of the wreck is the bottom hull and lower deck, offering divers of all experience levels the opportunity to float through the massive, open hull.

Ana Cecilia position

Name:	*Ana Cecilia*	**Last owner:**	Ernso Borgella
Type:	Cargo ship	**Sunk:**	July 13, 2016
Previous names:	n/a		
Length:	170ft (52m)		
Tonnage:	629grt		
Construction:	Halter Marine, USA 1972		

Just beyond *Amaryllis* lie the boulders of Habitat Corridors North, which help lead the way to the final wreck in the trek, the *China Barge* (also known as Research Team Reef). Little is known about this barge other than it sank in 1993. The artificial reef is an 80-foot-long (24.5-meter) flat-topped barge that is well encrusted in hard and soft corals as well as small sponges. The strength of the currents in this part of the coast means that most of the growth on the barge has a low profile, but the coverage is quite extensive. The barge offers shelter to many species of reef fish, including goliath grouper, which often hang out in the space between the sand and the underside of the bow of the barge.

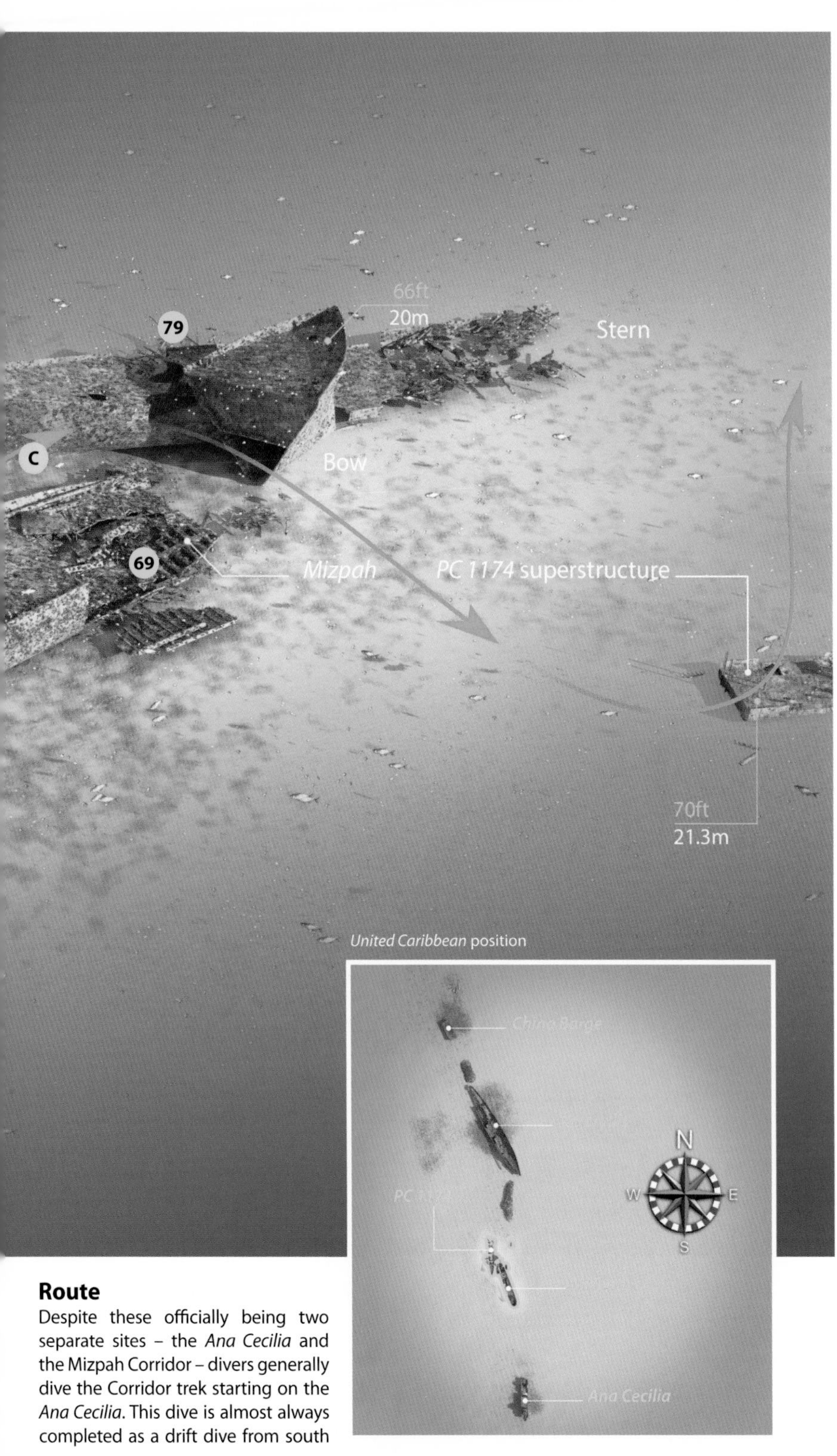

Route

Despite these officially being two separate sites – the *Ana Cecilia* and the Mizpah Corridor – divers generally dive the Corridor trek starting on the *Ana Cecilia*. This dive is almost always completed as a drift dive from south

Name:	*Mizpah*	**Construction:**	Newport News, Virginia, 1926
Type:	Yacht	**Last owner:**	Eugene Kinney
Previous names:	*Savarona, PY-29, Sequoia Allegro*	**Sunk:**	April 8, 1968
Length:	185ft (56.6m)		
Tonnage:	548grt		

Name:	*PC-1174*		Wisconsin, July 1943
Type:	Patrol craft	**Last owner:**	Private interests
Previous names:	*USS Fredonia*	**Sunk:**	April 9, 1968
Length:	165ft (50.5m)		
Tonnage:	432grt		
Construction:	Leathem D. Smith Shipyards,		

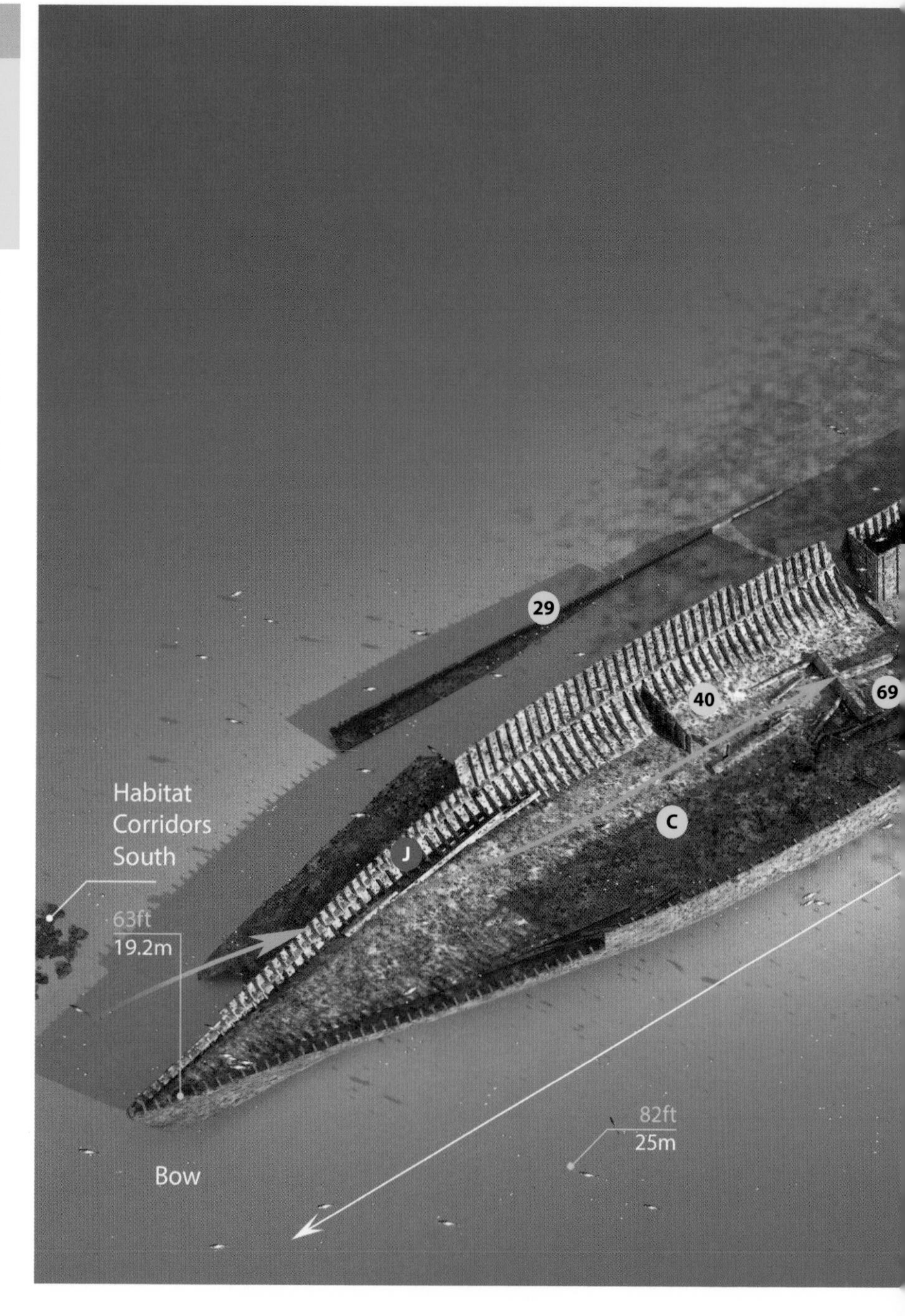

Name:	*Amaryllis*	**Construction:**	Burrard Drydock, Vancouver, 1945
Type:	Freighter	**Last owner:**	W. Coupland & Co, Cardiff, UK
Previous names:	*SS Cromwell Park, SS Harmac Vancouver*	**Sunk:**	August, 1968
Length:	440ft (135m)		
Tonnage:	7,147grt		

Habitat Corridors North
80ft
24.5m
Stern
53ft
16.2m
79
82ft
25m
440ft / 135m
W
N
S
E
Amaryllis position
China Barge
Amaryllis
N
W
E
S
Ana Cecilia

to north given the typically strong, prevailing currents along this part of the Florida coast. Most operators drop divers into the water just south of the *Ana Cecilia* so they can descend to the bottom as they drift to the north. A current is necessary for divers to have a chance of visiting the entire trek. Even so, most divers find that their air is a limiting factor by the time they reach the *Amaryllis*. Few make it all the way to the Habitat Corridors North and on to *China Barge*.

It is not worth rushing through the first few wrecks in the trek to have a chance at visiting the last ones, however. Divers should take their time

China Barge position

exploring at a leisurely pace. Most of the wrecks at this site have been on the seafloor for decades, meaning they have become well colonized with corals and sponges and host a diverse population of reef fishes.

Blue Heron Bridge (Phil Foster Snorkel Trail)

Difficulty ●○○
Current ●●○
Depth ●○○
Reef ★☆☆
Fauna ★★★

Access about 5 min from Riviera Beach
about 1 min from shore

Level Open Water

Location

Riviera Beach, Palm Beach County
GPS: 26°47′00.5″N, 80°02′33.9″W

Getting there

Blue Heron Bridge is a shore-accessible snorkel and dive site located in Phil Foster Park. The site gets its name from the Blue Heron Bridge, which spans the Intracoastal Waterway between Riviera Beach and Singer Island. There is plenty of free parking in the park, along with bathroom and outdoor shower facilities.

To get there, drive east on Blue Heron Boulevard. Turn left into the park on the small island that sits in the middle of the Intracoastal Waterway. The park's official address is 900 Blue Heron Boulevard, Riviera Beach, Florida 33404.

Access

This site is accessible to snorkelers and divers of all experience levels. There are two main access points to the southeast and southwest – on either side of the park's designated swim area. Divers are not allowed in the designated swim area.

The easiest access point is to the southwest, as divers can park near the playground structures in the southwest corner of the park, don their gear at their cars and walk the 100 feet (30 meters) to the beach, passing underneath the bridge structure. The park's featured snorkel trail, the Phil Foster Snorkel Trail, stretches from east to west, parallel to the beach 230 feet (70 meters) from shore. Most divers like to explore the bridge pilings to the east and west of the site as well. And given the shallowness of this site, there is little concern about running out of bottom time.

Blue Heron Bridge at dusk.

Matt9122/Shutterstock ©

Sea horses are found at Blue Heron Bridge.

Reef Smart Guides ©

It is important to pay attention to the tide charts given the strong currents that wrap around the island. Visibility also decreases with an outgoing tide. The best time to access this site is during the two-hour period that straddles high tide. Ideally, divers should aim to be putting on their dive gear about an hour before high tide, so it is slack water by the time they reach the bridge pilings. To find out the exact time for high tide at Blue Heron Bridge, look up tides for the Port of West Palm Beach (StationID# 8722588) on the NOAA tide charts (**Tidesandcurrents.noaa.gov**). Divers must use a dive flag and snorkelers need a flag if they intend to leave the guarded swim area.

The park is accessible from sunrise to sundown, but Blue Heron Bridge is also a very popular night diving location when the tide cycle is right. Divers require a permit to dive here at night and should call the Palm Beach parks and recreation department for more information (Tel: 561-963-6707). Alternatively, they can coordinate with Riviera Beach dive shops, some of whom are permitted to be in the park after dark on certain evenings.

Description

Sport Diver magazine voted Blue Heron Bridge one of the 50 best dive sites in the world and SCUBA Diving Magazine called it Florida's "best shore dive." The reason for its popularity is largely down to its easy accessibility and the incredible biodiversity found in the shadows of a heavily developed stretch of the Florida Intracoastal Waterway. Divers and snorkelers have the opportunity to encounter eagle rays, seahorses, octopuses, flying gurnards, frogfish and pipefish, among many others.

BLUE HERON BRIDGE (PHIL FOSTER SNORKEL TRAIL)

Other species commonly found at this site: I J 2 7 18 31 35 46 47 49 55 74 77 78 80 88

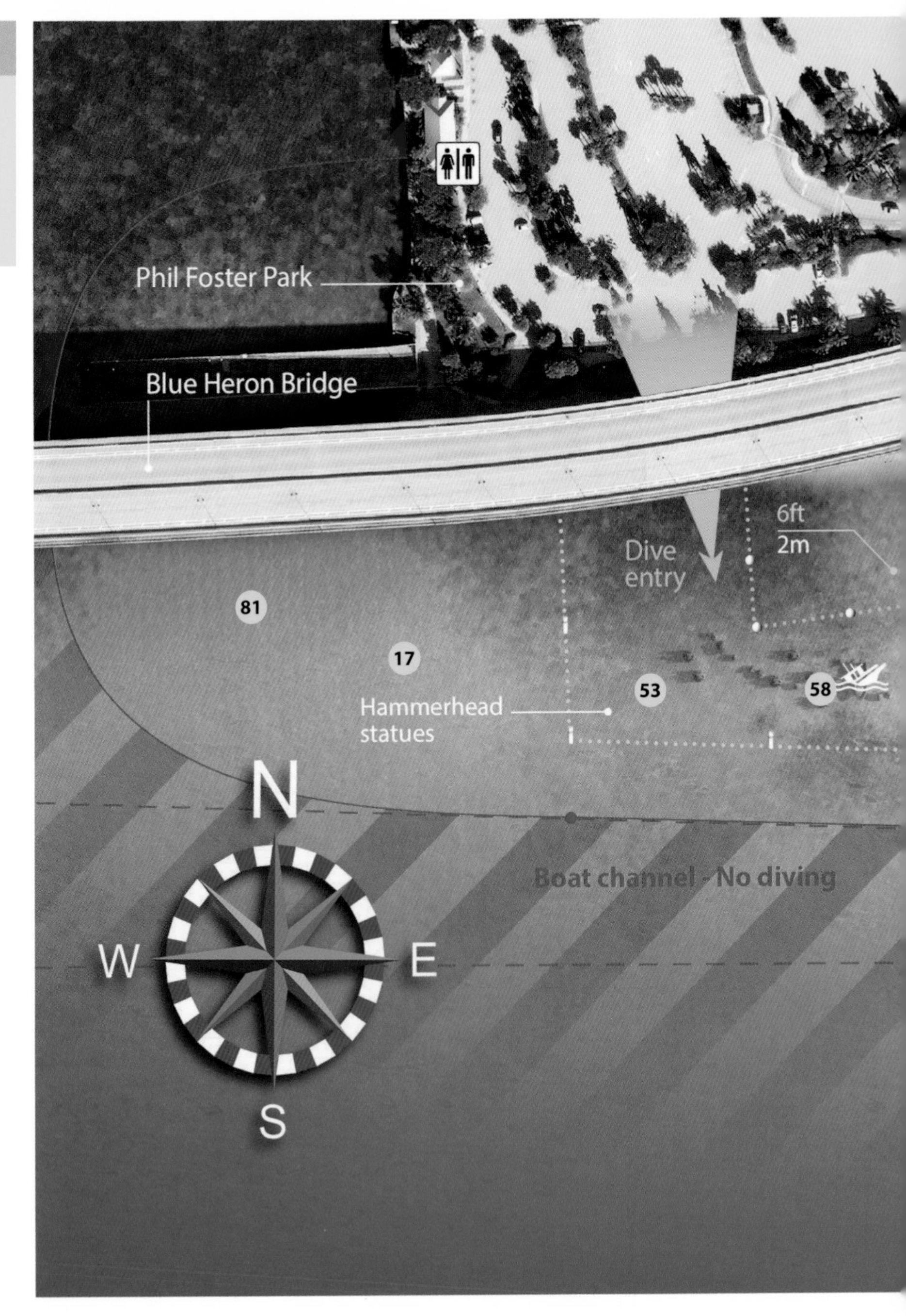

The site offers a range of habitats, including bridge pilings that sit to the west and east of the island. The 800-foot-long (244-meter) snorkel trail, which parallels the beach just outside of the guarded swim area, rests on a sandy bottom at a depth of 14 feet (4 meters). The trail consists of two concrete artificial reef designs, complex "Andrew Red Harris Foundation" reef modules and more simplistic concrete modules. There is a single wreck along with multiple shopping carts mixed in with the official snorkel trail.

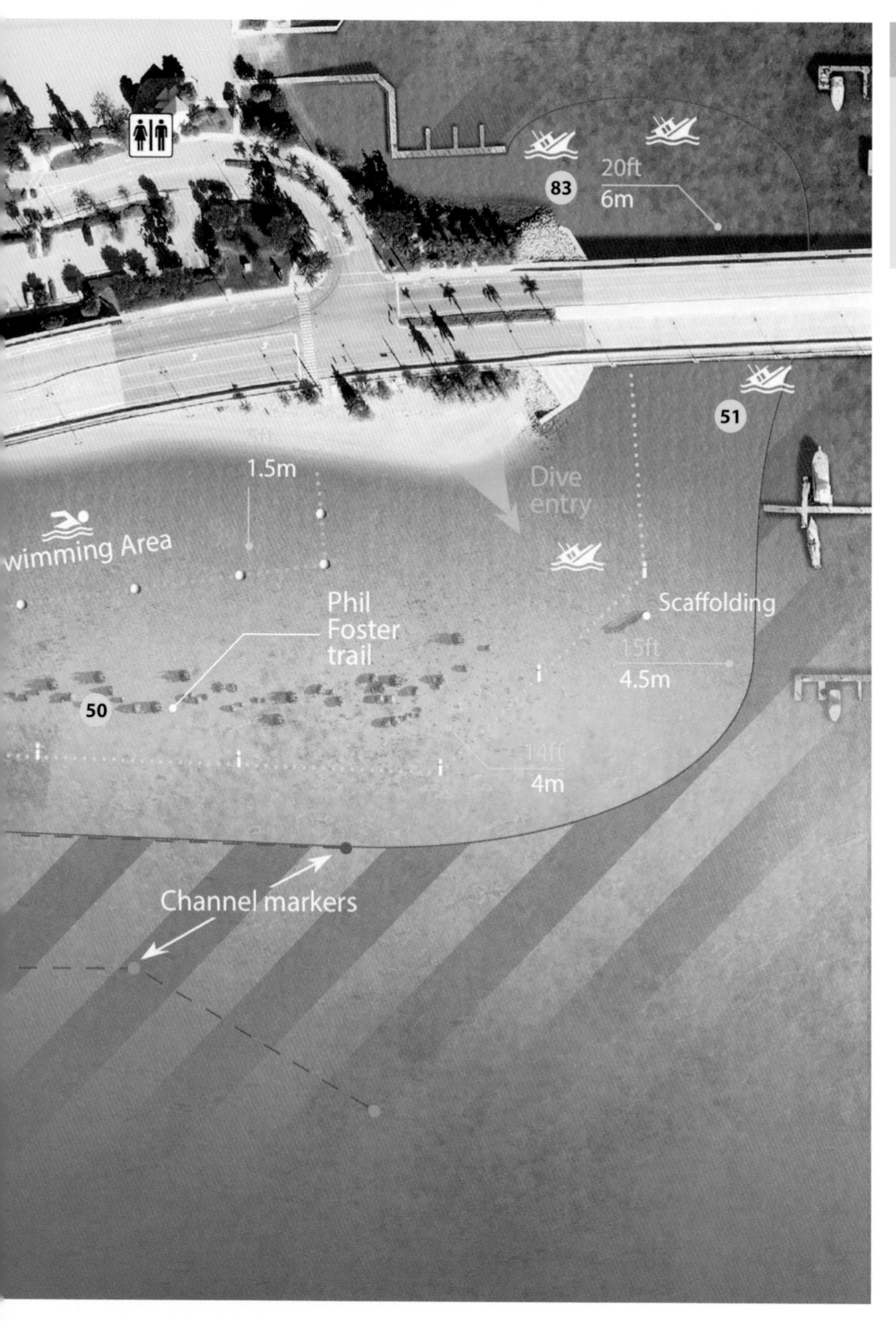

The broader site offers an eclectic mix of other artificial reef structures, including some scaffolding, small sculptures of hammerhead sharks located at the west end of the snorkel trail and several boat wrecks that can be found to the east, including one adjacent to the bridge pilings near the eastern channel.

A boat channel surrounds the island, representing an area that is off limits to divers and snorkelers; channel buoys mark this no-swim zone. Fishing takes place off the short fishing pier to the west

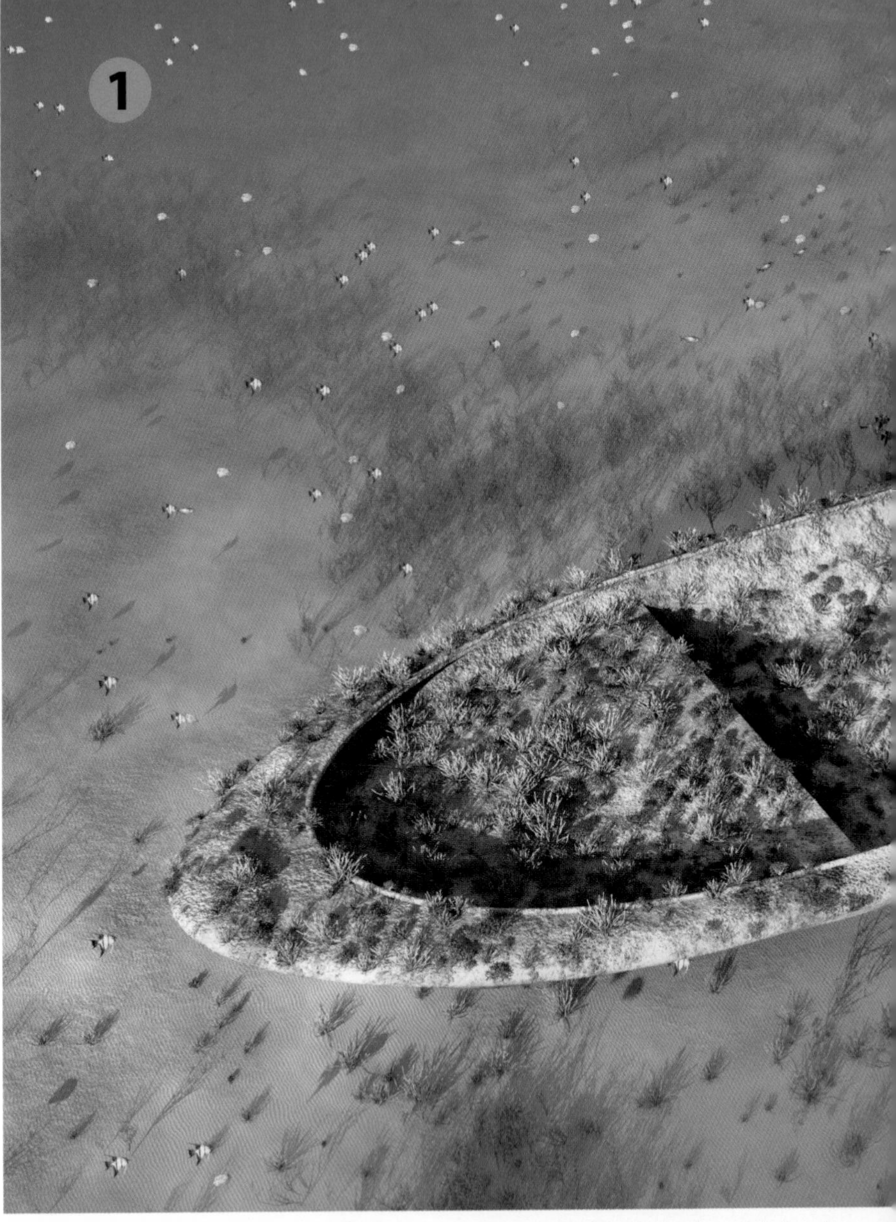

of the park, so it is important for divers and snorkelers to be aware of the potential for fishing lines and lures in the water in this area.

Route

A typical route involves entering the water on the beach to the west of the guarded swim area. Divers and snorkelers can access the reef by swimming directly out from shore. From here, they can either turn east to explore the length of the snorkel trail or west to catch the western end of the trail before heading northwest toward the bridge pilings. If they head east, they can continue past the end to the trail to the bridge pilings to the east of the island. Here they will spot the wreck of a sailboat that is wedged at the base of the pilings as well as the hull of a speedboat. Schools of Atlantic spadefish are often seen here, and divers should look

DID YOU KNOW?

For over a decade, the Florida Fish and Wildlife Conservation Commission (FWC) received numerous requests from the public to establish a "marine sanctuary" around the Blue Heron Bridge dive and snorkel area. In response to a rise in the "harvesting" of marine creatures around the bridge in 2017 and 2018, driven particularly by the aquarium trade, the FWC amended their Marine Life regulations to prohibit the collection of all marine life at this site as of April 1, 2019. The new rules do not affect the hook-and-line fishing that takes place from the fishing pier located on the western edge of the site.

closely for cryptic seahorses in this area. Be aware that the boat ramps are located on the eastern edge of the park, so the no-swim zone begins just beyond the bridge pilings. From here, divers can return along the trail to explore the western end of the site, paying particular attention to incredible biodiversity found both along the trail and under the bridge that connects the island to the mainland. Divers may see spotted eagle rays in this part of the site. The bridge pilings themselves are heavily encrusted with a mix

of sponges and hard corals, creating a mix of habitat for a variety of reef fish species.

SCIENTIFIC INSIGHT

Many artificial reefs are made out of old ships as well as other structures such as oil rigs, piles of concrete rubble or bridge pilings. However, research on what makes a good reef has helped identify certain shapes and characteristics that help enhance the ability of an artificial reef to create habitat for reef fish and corals. For instance, the size, density and shape of the holes in the artificial reef

will impact the species and types of reef organisms that choose to colonize it. Even the shape of the reef module itself impacts how the currents interact with the artificial reef, and thus how larvae of everything from corals to fishes will settle on it – and sediment that can harm developing corals. There are many different module designs out there, and yet many different reefs deploy some of the same designs. Next time you dive on an artificial reef, pay attention to the reef as well as the creatures that now call it home.

ECO TIP

When conducted responsibly and sustainably, aquariums can help the public develop a greater appreciation and understanding of coral reefs. But all too often, the aquarium trade involves the illegal and unsustainable poaching of organisms from living reefs.

Frogfish and seahorses are particularly lucrative targets in the aquarium trade because of their unique appearance. While frogfish are not themselves threatened, seahorse populations have declined across much of their distribution, in part due to harvesting for the aquarium trade. This was the main reason for the FWC's new law.

5

6

Spearman's barge

Difficulty
Current
Depth
Reef
Fauna

Access about 12 mins from Lake Worth Inlet

Level Open Water

Location

Riviera Beach, Palm Beach County
GPS: 26°46'57.2"N, 80°00'038.5"W

Getting there

Spearman's Barge lies 1.4 miles (2.3 kilometers) northeast of the Lake Worth Inlet. It requires a short boat ride of around 12 minutes to get there from the mouth of the inlet. It is only accessible by boat due to its distance from shore and the best way to reach it is through one of the local dive operators who use the Lake Worth Inlet.

Access

There is no permanent mooring buoy on the wreck, so local operators often put divers in the water slightly up current from the site. Currents typically flow north along the coast. *Spearman's Barge* rests just off shore of the popular Mizpah Corridor. Visibility is generally good here. The wreck of the barge is surrounded by a mix of sand and reef habitat at a depth of just 71 feet (21.5 meters) and is accessible to divers of all experience levels. The use of a surface marker buoy is required at this site.

Description

Spearman's Barge is also sometimes referred to as *Colson's Barge*. The 150-foot-long (45.5-meter) deck barge was sunk in 1978 to form an artificial reef. It bears the name of its owner at the time of its sinking, Robert Spearman. Reportedly an avid lobsterman and spearfisherman, Spearman ran a small dive shop that grew to become a large company operating under the name of Spearman Marine Construction. He sank the barge as an artificial reef to help attract the large fish he liked to spear.

The wreck offers the chance to see large schools of grunts, as well as sergeant majors, butterflyfish and angelfish. Parrotfish and surgeonfish forage on the algae that covers the hull of the barge, while honeycomb cowfish and sharpnose puffers flit about the site. After

Schooling snapper are a common sight on this wreck.

DID YOU KNOW?

The story of the barge is relatively straightforward, but the stories that surround Spearman himself are far more interesting. On November 16, 1985, Anita Spearman, Robert Spearman's wife, was found beaten to death at the couple's Palm Beach Garden home.

The police investigation discovered that Spearman had answered an advertisement in Soldier of Fortune magazine for a "gun for hire." The ad was placed by a Tennessee

so many years underwater, the artificial reef is heavily colonized by hard and soft corals, and it has helped create complex habitat that complements the surrounding natural reef. A resident goliath grouper often greets divers as they approach the wreck.

Route

Most divers are dropped in the water to the south of the wreck. They descend to the bottom to drift with the current until reaching the barge. Under low-current conditions, divers can spend plenty of time touring the barge and the various openings that may contain surprise macrofauna, including turtles, before spreading out to explore the surrounding reef. The current is often strong at this site however, which means divers usually only have time for a single pass over the wreck before being carried into the reef habitat.

The barge sits to the east of a long reef tract that

Eric Carlander/Shutterstock ©

nightclub owner by the name of Richard Savage. Spearman, the nightclub owner and his hired hitman, Sean Doutre, were all tried and convicted for crimes tied to the murder.

Spearman received a life sentence plus an additional 20-year term to be served at Zephyrhills Correctional Institution. He hatched a plan to escape, paying what turned out to be two undercover law enforcement agents a $4,700 down-payment on a $54,700 contract to break him out of prison. His plan? The agents would fly into the central yard of the prison facility in a helicopter with two machine guns and four hand grenades.

Three days after the foiled escape attempt, guards found Spearman's body – he reportedly committed suicide rather than face his soon-to-be-extended prison sentence. In another twist, Spearman apparently left one million dollars to his ex-wife shortly before his trial. After his death, the ex-wife maintained the money was an outright gift, and refused to turn the money back over to the state.

runs parallel to the shore. After exploring the wreck, most divers drift north with the current, exploring the complex habitat on the east side of the reef tract.

SPEARMAN'S BARGE

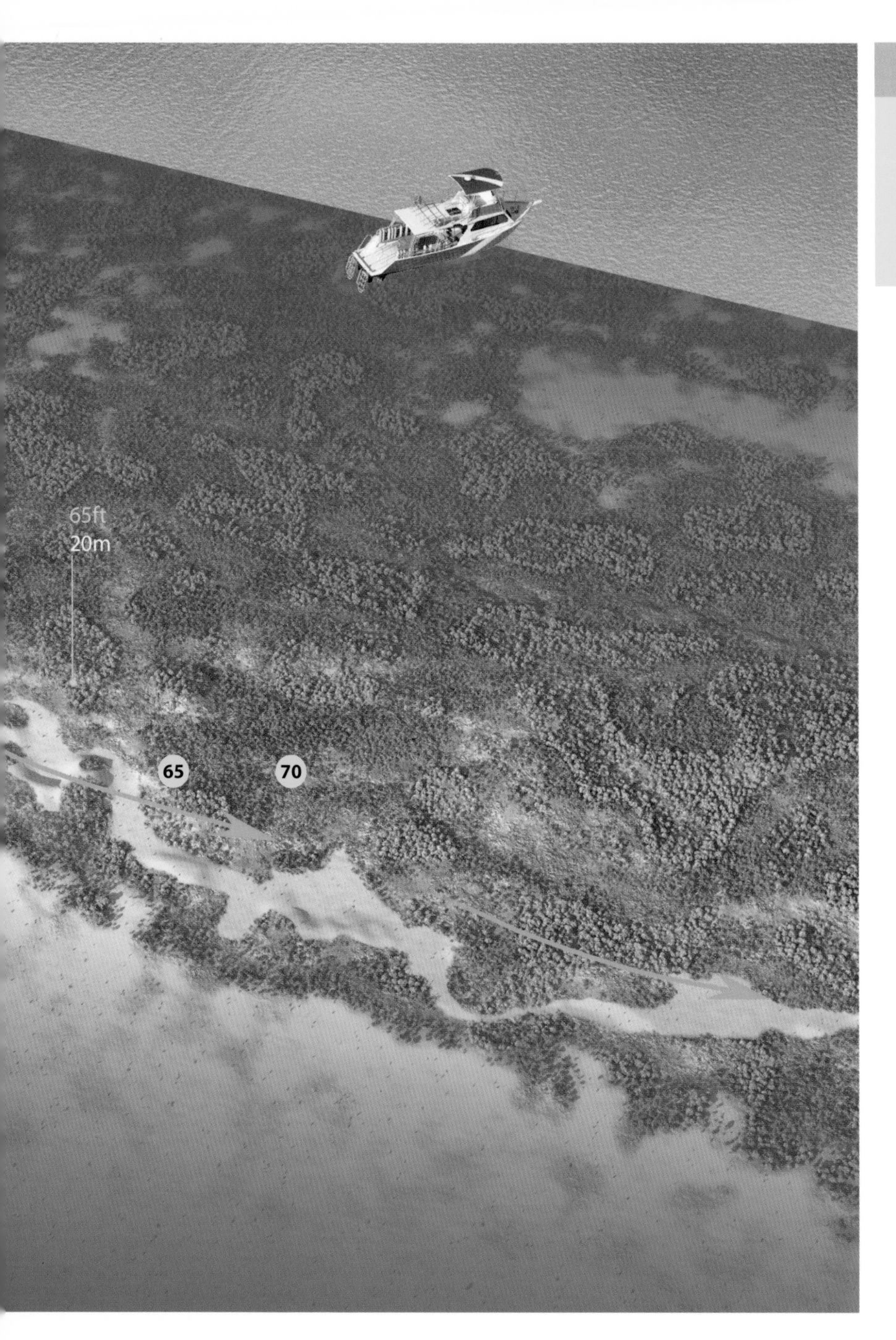

Other species commonly found at this site: K 2 8 12 14 18 29 30 31 33 36 48 63 66 67 68

Name:	*Spearman's Barge*	**Last owner:**	Spearman Marine Construction
Type:	Deck barge	**Sunk:**	1978
Previous names:	n/a		
Length:	150ft (45.5m)		
Tonnage:	Unknown		
Construction:	Unknown		

150ft/ 46m
65
39
72ft
22m

Eidsvag Triangle

Difficulty ●●●
Current ●●○
Depth ●●○
Reef ★☆☆
Fauna ★★☆

Access about 8 mins from Lake Worth Inlet

Level Open Water

Location

Riviera Beach, Palm Beach County
GPS (*Eidsvag*): 26°45′56.1″N, 80°0′44.3″W

Getting there

The *Eidsvag* Triangle, sometimes referred to as the Palm Beach Triangle, includes three wrecks that form a triangle, located 1.25 miles (2 kilometers) off shore of Riviera Beach – about an 8-minute boat ride east of the Lake Worth Inlet. The best way to reach the wreck trek is through one of the area's local dive operators.

Access

There is no permanent mooring buoy on any of the wrecks at this site, so local operators often put divers in the water slightly up current of the site so they have time to descend toward the bottom and allow the currents to push them to the wrecks. Divers typically explore the trek as a drift dive from south to north, following the prevailing currents. Visibility is generally good on this site but currents can be strong, which may make visiting the smallest artificial reef in the trek, located off to the east of the main wrecks, more challenging. This dive is most suited to those with experience in deeper water and with stronger currents, particularly due to the proximity of the site to the inlet and the heavy boat traffic.

Description

The *Eidsvag* Triangle gets its name from the main wreck at this site, the 150-foot-long (45.5-meter) *Eidsvag*. The British-built coastal freighter was deliberately sunk as an artificial reef in 1985. The

SCIENTIFIC INSIGHT?

Humans have been dropping artificial reef structures into the water dating back to 250 BC. Originally these were used as a way to defend important ports against attack by pirates. But people soon realized these structures attracted fish. By the 1800s, artificial reefs were routinely being used around the world to improve fish catches, and much later, to create habitat for corals. Unfortunately, not all artificial reef structures are created equal. In general, concrete and thick steel – such as that used to build ships – offer corals a good surface on which to settle. By comparison, rubber and thin metal – such as the body of a car – provide less suitable habitat. The simple reason is that seawater is a harsh environment for metals, and the thin metal used in a car chassis rusts away too quickly.

other wrecks in the triangle include *Murphy's Barge*, a 120-foot-long (36.5-meter) flat-bottom barge, and the remains of a 1967 Rolls Royce Silver Shadow donated by a Palm Beach County hairdresser named Greg Hauptner. The three artificial reefs sit near one another on a sand and rubble seafloor that bottoms out at a depth between 82 and 87 feet (25 and 26.5 meters).

Murphy's Barge (also known as *Phillip's Barge*) is the southernmost wreck in the triangle. The barge was one of many donated over the years by the Murphy Construction Company, a local marine construction company and supporter of the Palm Beach County artificial reef program.

This particular barge was the last of the three artificial reefs deployed by the county to form the trek. It was deployed along with piles of concrete posts that have since settled onto the surrounding seabed.

The hull of the barge is open in some places but offers no opportunity for penetration. Divers report regularly seeing a green moray eel on the wreck. The hull is only partly colonized by corals and sponges and is mostly covered in algae.

The bow of the *Eidsvag* sits just over 10 feet (3.5 meters) to the northwest of the barge. The *Eidsvag* was a Honduran-flagged coastal freighter built in 1941. The county scuttled the vessel as an artificial reef in December 1985 using 110 pounds of explosives. It was the first large-scale contribution to the county's nascent artificial reef program.

The freighter goes by *Eidsvag*, *Owens* or even *Eidsvag and Owens*, depending on who you ask. The County Commissioner at the time, Jerry Owens, was a driving force in acquiring the freighter at a live auction in Miami for $9,500. His name has stuck with the artificial reef as a result.

After the county deployed the artificial reef, the first divers to inspect the wreck as it lay on the seafloor reported that not only did the freighter just barely miss the vintage Rolls Royce sunk in the same location just months earlier, but the area was ringed with dead fish, presumably killed from the concussive force of the blast.

Wrecks provide complex habitat for angelfish and other reef creatures.

Peter Leahy ©

Plenty of reef fish species have recolonized the wreck in the intervening years. Divers will encounter black margates, grey snapper, angelfish and schools of grunts on the wreck, as well as Atlantic spadefish and bar jacks in the water column above it. The decades underwater have not been kind to the wreck, however. As with many coastal freighters sunk as artificial reefs, the wreck now consists of separate bow and stern sections bracketing a flattened debris field. The stern rests on its port side more than 150 feet (45.5 meters) to the north. Its surface has been heavily colonized by both hard and soft corals as well as sponges. Goliath grouper are a common

Other species commonly found at this site: 1 2 3 4 6 11 14 20 26 28 33 35 42 44 47 65

sight on this wreck, hiding in the shadows and finding shelter from the current. There are no penetration opportunities on the *Eidsvag* as the structure has deteriorated significantly during its time underwater.

The final "wreck" in the triangle – and the first of the three artificial reefs to be deployed – lies nearly 40 feet (12 meters) northeast of the barge. All that is left of the vintage 1967 Rolls Royce that Greg Hauptner helped roll off a barge and into the water in September 1985 are the wheel wells, engine block, steering wheel and rear chassis of the car. While it is easy to recognize that this

was once a car, long gone are any identifiable characteristics that might give away its high-class origins.

An avid diver, Greg Hauptner was one of many in the Palm Beach diving community disappointed when neighboring Broward County purchased the Mercedes I to sink as an artificial reef to the south. The vessel had earned celebrity status while beached on a stretch of Palm Beach County's affluent coastline. As reported in the news at the time, the hairdresser-to-the-stars got the idea after hearing Jerry Owens on the radio talking about reviving the county's artificial reef

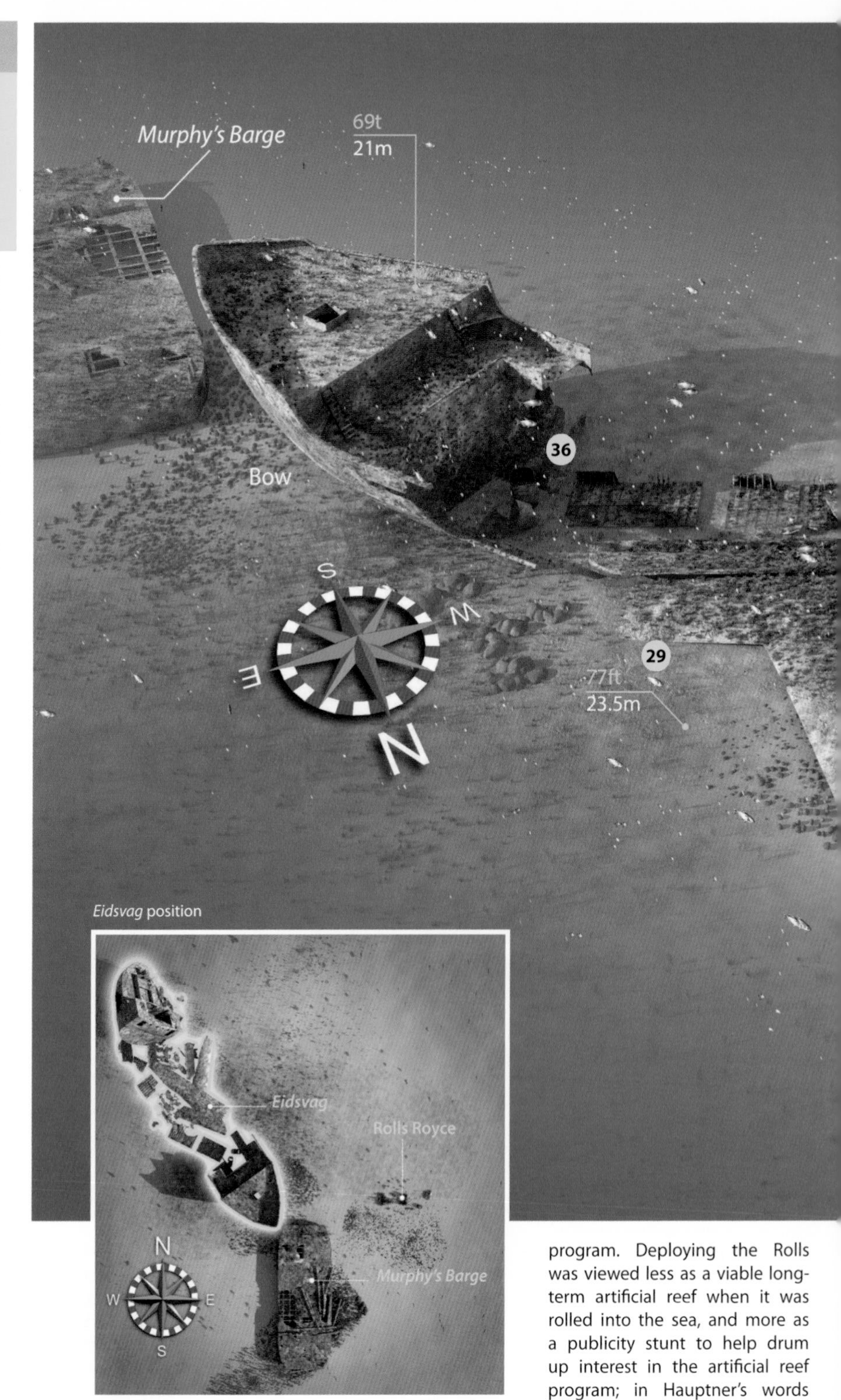

program. Deploying the Rolls was viewed less as a viable long-term artificial reef when it was rolled into the sea, and more as a publicity stunt to help drum up interest in the artificial reef program; in Hauptner's words

at the time "Palm Beach loses a Mercedes and gains a Rolls." And it worked – the story received coverage in the U.S. and in overseas media markets and did indeed help kick off an artificial reef program in Palm Beach County that remains active and effective today.

Route

Divers typically start out on the southern end of the triangle given the prevailing currents typically run toward the north. In mild to moderate currents, divers can easily make their way around the triangle, visiting all three wrecks

in a clockwise direction, from the barge to the freighter and then heading back up-current to the Rolls Royce. In stronger currents, however, divers may be forced to skip the car in favor of spending more time on the barge and the remains of the *Eidsvag*. While the Rolls represents an interesting curio, it does not support much in the way of marine life so there is little reason to dedicate too much time to that part of the trek.

Name:	*Eidsvag*	**Last owner:**	Unknown
Type:	Freighter	**Sunk:**	December 16, 1985
Previous names:	*USS Owens*		
Length:	150ft (45.5m)		
Tonnage:	315grt		
Construction:	Richards Ironworks, U.K., 1941		

Stern
71ft
21.5m
M
77ft
23.5m
29
Murphy's Barge position
Eidsvag
Rolls Royce
Murphy's Barge
N
W
E
S

Name:	*Murphy's Barge*	**Last owner:**	Murphy Construction Company
Type:	Barge		
Previous names:	*Phillip's Barge*	**Sunk:**	1985
Length:	120ft (36.5m)		
Tonnage:	Unknown		
Construction:	Unknown		

75ft
23m
30
75
46
Rolls Royce position
Eidsvag
Rolls Royce
Murphy's Barge
N
W
E
S

Name:	Rolls Royce	**Last owner:**	Greg Hauptner
Type:	Car	**Sunk:**	September 23, 1985
Previous names:	n/a		
Length:	16.5ft (5m)		
Tonnage:	Unknown		
Construction:	1967		

Breakers North & Turtle Mound

Access  about 25 mins from Lake Worth Inlet

Level Open Water

Location

West Palm Beach, Palm Beach County
GPS: 26°42'46.9"N, 80°00'18.6"W

Getting there

Breakers North and Turtle Mound represent a section of reef ledge that runs roughly parallel to the coastline just under one mile (1.5 kilometers) off shore from West Palm Beach.

Many sea turtle species can be found on Turtle Mound.

They sit about 4 miles (6.5 kilometers) south of the Lake Worth Inlet. It is only accessible by boat due to its distance from shore. The best way to reach Breakers North and Turtle Mound is through one of the local dive operators.

DID YOU KNOW?

The Breakers Hotel has a rich history. The Italian Renaissance-style hotel was originally known as The Palm Beach Inn when it first opened in 1896. The name changed to The Breakers when guests at its sister resort, the Royal Poinciana Hotel located a half mile away on the Lake Worth Lagoon, started to request rooms "down by the breakers," referring to the waves on the Atlantic coast.

Henry Flagler, a founding partner of Standard Oil, built both hotels after seeing the state's potential as a vacation destination. He invested large sums of money in the area, including railway infrastructure in the form of the Florida East Coast Railway. He built many hotels including The Ponce de Leon, Alcazar and Royal Palm hotels, although The Breakers was the most famous of these.

In the early 20th century, the hotel hosted a who's who of the Gilded Age's rich and famous, including John D. Rockefeller, Andrew Carnegie, J.P. Morgan, J.C. Penney, the Vanderbilts, the Astors and various U.S. presidents and European royalty. The hotel twice went up in flames but was rebuilt both times, and still stands tall and proud over a century after it first opened.

Access

This site is accessible to divers of all experience levels, and most divers explore it as a drift dive. Visibility is generally good, but currents can be moderate to strong, which can help divers explore more of the site on their dive.

The reef top is interesting to explore in some places, but the small ledge on the western edge of the reef, adjacent to the sand, is the more rewarding dive. This area supports the highest species density and diversity. There are no mooring buoys at this site, and divers should bring a surface marker buoy with them when they dive.

Description

Breakers North and Turtle Mound are two distinct reef areas that are often explored as part of the same dive. Breakers North is a contiguous ledge that runs parallel to shore for approximately 2 miles (3 kilometers). It gets its name from the Breakers Hotel, which is located on the shore adjacent to the reef.

Turtle Mound is a large dome-shaped reef located at the northern end of Breakers North and is separated from the main ledge by a stretch of sand approximately 100 feet (30.5 meters) across. The two sites are widely regarded as among the best dive sites in Palm Beach County, which makes them particularly popular with local operators.

Zachary R. Nolan ©

Divers usually explore the shallower west side of the reef ledge, which rises about 10 feet (3 meters) from the seabed at a depth of 60 feet (18 meters). The ledge is frequently broken up by fingers of reef that extend out into the sand, toward shore. The ledge itself is undercut in many places and divers may find numerous species sheltering beneath the overhangs, including lobsters, nurse sharks, moray eels and goliath grouper. Divers may also see yellowhead jawfish in the sand beside the reef. The top of the reef, which crests at a depth of around 50 feet (15 meters), is relatively flat and dominated by soft corals and sponges.

Divers will find an amphitheater toward the northern end of Breakers North that shelters soft sand and is a favorite hiding spot for southern stingrays. A large, coral-encrusted, statue of Neptune once stood watch over the reef here, but it has since toppled and is now buried in the sand, leaving just the concrete pedestal behind.

At the northern end of the amphitheater, the ledge is deeply undercut at a spot know as "the cave," where divers often encounter goliath grouper and large schools of grey snapper, grunts and porkfish. With so much going on along this stretch of ledge, it can be hard for divers to know where to look, whether at sand, reef or even the water column above the reef where barracuda and schools of Atlantic

spadefish may be found. Divers often get a pleasant surprise when they reach Turtle Mound, which is famous for its abundance of sea turtles. Hawksbill, green and loggerhead sea turtles are often found feeding, and sometimes sleeping, in and around this area, particularly during the spring and summer months.

Route

There is often a moderate current at Breakers North that runs north toward Turtle Mound. As such, these sites are often explored as a drift dive. Most divers are dropped at a location known as Fourth Window, because it literally lines up with the forth window on the north section of The Breakers Hotel.

Divers drift with the current to the north. They may notice a telecommunications cable that runs perpendicular to their route, directly across the reef. The cable runs from West Palm Beach to Brandie Point on Grand Bahama Island – a distance of 84 miles (135 kilometers) – and then continues on to Goodman Bay in New Providence for another 171 miles (276 kilometers).

The reef in the southern part of Breakers is not as spectacular as its northern neighbor. There is still plenty to see, however, including morays, Bermuda chub and schools of snapper. The amphitheater and cave are where divers spend most of their time, but divers should conserve some air to reach Turtle Mound.

Turtle Mound is as much as 300 feet (91.5 meters) across, rising steeply on the south edge and sloping more gently into the sand on the northern side. A smaller mound is located slightly to the south of Turtle Mound and due west of the cave which can assist in navigation.

Other species commonly found at this site:

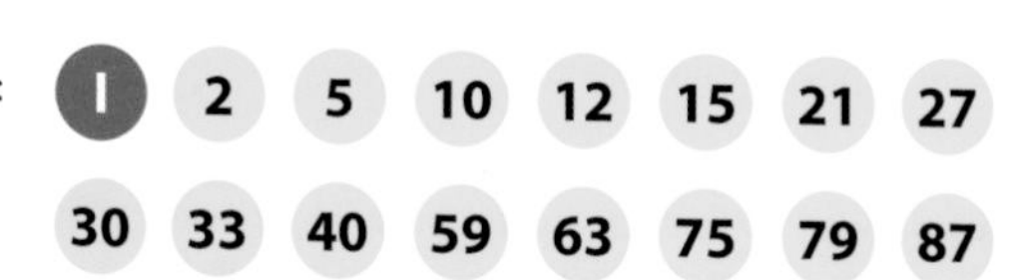

Trench

Difficulty ●○○
Current ●●○
Depth ●●○
Reef ★★☆
Fauna ★★★

Access about 28 mins from Lake Worth Inlet

Level Open Water

Location

West Palm Beach, Palm Beach County
GPS: 26°42'18.5"N, 80°00'59.4"W

Getting there

Trench is a section of reef ledge located between the Breakers and Teardrop dive sites. It forms part of what is sometimes referred to as the Breaker's Reef system. This stretch reef runs parallel to the coastline where the Breaker's hotel can be found – hence the name. Trench is located about 1 mile (1.5 kilometers) off shore from West Palm Beach and about 4.5 miles (7.3 kilometers) south of the Lake Worth Inlet. It is only accessible by boat due to its distance from shore. The best way to reach Trench is through one of the local dive operators.

Access

This site is accessible to divers of all experience levels. Most divers explore Trench as a drift dive. Visibility is generally good, but currents can be moderate to strong, which can help divers explore a greater amount of the site on their dive. There are no mooring buoys at this site; divers should bring a surface marker buoy with them when they dive here.

Description

Trench is sometimes referred to as Breakers South, Moray Alley or Outfall Trench, referring to the site's main feature – a 12-foot-wide (3-meter) artificial trench that cuts through the reef at roughly 5 feet (1.5 meters) below the top of the reef. The trench runs for about 200 feet (61 meters) from west to east and bottoms out at an average depth of around 60 feet (18 meters). Although it appears empty, the trench houses an outfall pipe buried beneath the sediment at the bottom. The outfall pipe does not function often these days, other than to discharge ground water from land during times of heavy rainfall, such as during the hurricane season.

The trench typically shelters a wonderful assortment of reef creatures, including nurse sharks, stingrays, lobsters, yellowhead jawfish and an unusually high density of moray eels, which is how the site came to be named Moray Alley by some operators. Divers often spend most of their bottom time exploring the trench, which also provides shelter from the current.

Just over halfway along the length of the trench from its western end, divers will see a pile of boulders and sand bags that have become incorporated into the reef. Beyond these bags the channel starts to slope into deeper water toward the seaward side of the reef.

SCIENTIFIC INSIGHT

Most species of reef fish swallow their prey by creating a negative pressure space in the throat, which pulls the prey toward the stomach. However, the unique physiology of moray eels makes them unable to accomplish this feat. Instead, they have an interesting adaptation that helps them swallow prey: pharyngeal jaws. This highly maneuverable set of jaws, located just behind their primary set of jaws, launches forward, bites down on the prey and then pulls it down the throat. The moray's pharyngeal jaws were only discovered in 2007, by scientists at the University of California, Davis.

The reef habitat to the south of the trench is generally of lower quality than what is found to the north. It supports smaller colonies of hard and soft corals as well as numerous barrel sponges, several of which have grown to around 5 feet (1.5 meters) in size. Cleaning stations are fairly common here, and schools of Bermuda chub, porkfish and grunts are often seen cruising the reef. The reef quality improves immediately to the north of the trench in an area known as Dive-O-Rama, where lemon sharks are frequently spotted. This part of the dive site is also a hotspot for small cryptic species, such as frogfish, flamingo tongues and cleaner shrimp.

Route

Divers are usually dropped about 1,000 feet (305 meters) to the south of the trench in an area known as Flower Gardens. As currents generally flow from south to north along this part of the coast, divers usually have plenty of time to descend to the reef and drift northwards before encountering the trench.

Divers may notice several telecommunications cables running perpendicular to the reef at Flower Gardens, depending on how far south they are dropped by their dive operator. These cables supply the nearby Bahamas islands. The

John A. Anderson/Shutterstock ©

The yellowhead jawfish is one of many interesting species common at Trench.

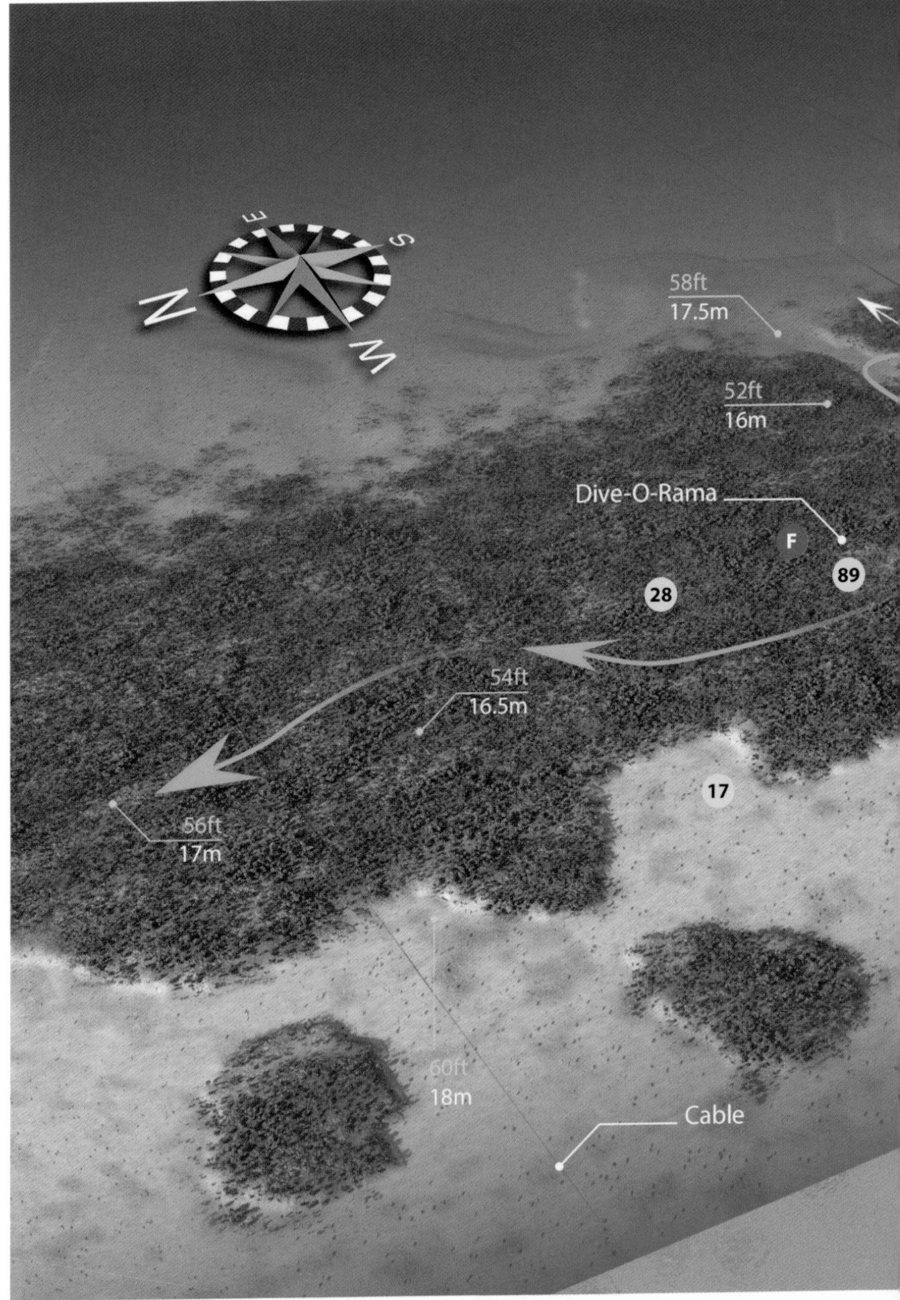

trench is located about 175 feet (53 meters) after the last of these cables.

The typical route for this dive takes divers into the trench. From there, most divers swim along the length of the trench, staying out of the prevailing current. Once they reach the coral encrusted boulders and sand bags located just past the halfway point, they turn back along the trench before crossing over to the reef that sits to the north, an area known as Dive-O-Rama. By heading northwest after Dive-O-Rama, divers can often locate yellowhead jawfish in the sand and rubble areas along the western edge of the reef, as well as the occasional midnight and blue parrotfish.

Other species commonly found at this site:

Teardrop

Difficulty ● ○ ○
Current ● ● ○
Depth ● ● ○
Reef ★★☆
Fauna ★★★

Access about 32 mins from Lake Worth Inlet

Level Open Water

Location

West Palm Beach, Palm Beach County
GPS: 26°41′17.4″N, 80°01′06.0″W

Getting there

Teardrop Reef is located on a section of reef ledge that runs roughly parallel to the coastline just under 1 mile (1.5 kilometers) off shore of West Palm Beach. The reef sits nearly 5.8 miles (9.4 kilometers) south of the Lake Worth Inlet and right between the dive sites named Bath and Tennis, and Flower Gardens. It is only accessible by boat due to its distance from shore. The best way to reach Teardrop is through one of the local dive operators that uses the Lake Worth Inlet.

Access

This site is accessible to divers of all experience levels. Most divers explore Teardrop as a drift dive, usually running from north to south. The reef top is interesting to explore in places, but the western edge of the reef is more rewarding and contains the highest species density and diversity, particularly around the feature known as Ron's Rock. This feature is located about 20 to 25 minutes into the dive. Visibility is generally good, but currents can be moderate to strong, which can help divers explore a greater amount of the site on their dive. There are no mooring buoys at this site; divers should bring a surface marker buoy with them when they dive here.

Description

Teardrop Reef also sometimes goes by the name Ron's Reef or Ron's Rock – the name applies to a key feature of the site located just before the sand gap between Teardrop and Middle Earth. The reef itself begins as a contiguous bank reef that runs parallel to shore. The top of the reef is relatively flat with an average depth of about 50 feet (15 meters), and it supports numerous small, hard and soft corals as well as medium-sized barrel sponges.

The east side of the reef is deeper and consists of the spur and groove features that are common on the seaward side of reefs throughout Palm Beach County. Divers most commonly explore the west side of the reef, however – the side that faces shore. This part of the reef leads directly to Ron's Rock and is generally more suitable for divers of all experience levels.

The west side of the reef starts out as a relatively nondescript scattering of patch reefs that form into two distinct ledges, each approximately 3 feet (1 meter) in elevation. As divers continue to head north, these two ledges move closer together. A clear sand-reef boundary exists to the west, at a depth of about 56 feet (17 meters), with a reef plateau at about 52 feet (16 meters) located

RELAX & RECHARGE

Havana has been serving arguably the best Cuban food in Palm Beach County for over a quarter century. Located in West Palm Beach on the Dixie Highway, Havana has a wide selection of Cuban favorites like ropa vieja and fricasé de pollo and as well some incredible sandwiches if you need your food to go. Havana has a great mix of quality food, reasonable prices, big portions and friendly staff. The restaurant is open seven days a week until late in the evening; if you are a night owl, Havana has you covered with their 24-hour walk-up window. Visit: **Havanacubanfood.com**

between the two ledges, and a reef top at about 48 feet (14.5 meters). Closer to the northern end of the contiguous reef, the two ledges fuse to become a single, much more pronounced cliff averaging a height of 10 feet (3 meters).

The cliff cuts sharply to the east around a headland into a small cove consisting of a collapsed cliff, where several large boulders have sheared off the ledge and formed caves and crevices packed with marine life. This area, which is known as Ron's Rock, has the highest diversity at Teardrop. Large schools of snapper, grunts and parrotfish dart among the rocks while lobsters are often found in the reef cracks. Sea turtles, goliath grouper and nurse sharks are occasionally spotted in the undercut areas of the reef. Ron's Rock is also notable for the large number of cleaning stations and juvenile fish that congregate here, as well as predators, including barracuda and even hammerhead sharks.

This section of north-south reef terminates at Ron's Rock. The central portion of the reef bank slopes gently down toward the sand and is covered in soft corals. As is the case along the full length of the reef, the depth gradually increases toward the eastern side, eventually connecting with the spur and groove habitat of the seaward side. North of Ron's Rock, open sand extends across what many operators refer to as

VisionDive/Shutterstock ©

Nurse sharks are occasionally spotted in the Ron's Rock area of Teardrop.

a 100-foot (91.5 meter) "leap of faith." Here, the sand supports a lot of life, including stingrays, garden eels, yellowhead jawfish and porgies. The next section of reef, known as Middle Earth or simply East-West Ledge, rises out of the sand abruptly, reaching a depth of about 48 feet (14.5 meters) before disappearing just as rapidly back into the sand.

Route

Teardrop is most often explored as a drift dive from south to north along with the prevailing currents. Most divers follow the western edge of the reef to Ron's Rock. The small cove and cliff provide some shelter from the current, which can allow divers to spend extra time here observing the diversity of marine life.

The jump to Middle Earth can require navigation skills if divers wish to explore the more interesting western edge of the outcrop, but the reef is wide enough that

divers are unlikely to miss it as they head north. Most divers end their dive at Middle Earth, but those using nitrox or with good air consumption rates may be able to cross the next stretch of sand and reach the southern tip of Flower Gardens Reef to the north.

Other species commonly found at this site: **1** **5** **8** **10** **11** **12** **21** **22** **25** **26** **27** **36** **45** **63** **64** **76**

Paul's Reef

Difficulty ●○○
Current ●●○
Depth ●●○
Reef ★★☆
Fauna ★★☆

Access about 34 mins from Lake Worth Inlet

Level Open Water

Location

West Palm Beach, Palm Beach County
GPS: 26°39'00.0"N, 80°01'02.0"W

Getting there

Paul's Reef is a section of reef ledge that runs roughly parallel to the coastline just 1.2 miles (2 kilometers) off shore of West Palm Beach. Paul's Reef sits about 7.5 miles (12 kilometers) north of the Boynton Inlet and about 8.5 miles (13.7 kilometers) south of the Lake Worth Inlet. It is only accessible by boat due to its distance from shore. The best way to reach this site is through one of the local dive operators using the Lake Worth Inlet.

Access

This site is accessible to divers of all experience levels. Most divers explore Paul's Reef as a drift dive. Visibility is generally good, but currents can be moderate to strong; a strong current can help divers explore more of the ledge on their dive. The reef top is interesting to explore in some places, but the small ledge that borders the sandy seabed on the western edge of the reef makes for a more rewarding dive since it contains a higher species density and diversity than the crest. There are no mooring buoys at this site; divers should bring a surface marker buoy with them when they dive.

Description

Paul's Reef is a long section of the reef tract that parallels the shoreline. It is about 250 feet (76 meters) wide, and most divers focus on the western side of the reef, which starts in the south as a double ledge system. Both ledges are about 3 feet (1 meter) in height. The first ledge sits adjacent to the sandy seafloor at a depth of about 55 feet (17 meters). The second ledge is shallower and located slightly east of the first ledge. The two ledges start about 75 feet (23 meters) apart, but that distance gradually narrows as the reef extends northward. They eventually join, forming a single tall and steep ledge that is undercut in places. Goliath grouper, nurse sharks and sea turtles are commonly found sheltering in these undercut areas.

Southern stingrays are common in the sandy areas of Paul's Reef.

The reef top is scattered with soft and hard corals and small barrel sponges. Large schools of tomtates and other grunts move throughout the area. Numerous shallow channels and recesses are cut into the reef, which often contain lobsters and crabs.

As divers drift north, the reef begins to bend noticeably to the northeast and the ledge becomes even taller and more undercut, reaching a height of about 12 feet (3.5 meters). In several places, parts of the ledge have collapsed entirely, leaving boulders in the sand and exposing channels that run back into the reef structure. Numerous species of parrotfish can be found on the reef, including midnight, blue and rainbow.

The ledge abruptly ends at a point called Sloan's Curve, where it bends sharply back toward the east. To the north and slightly to the west, divers will see that the reef tract continues as several isolated patches ranging from 100 to 300 feet (30 to 90 meters) in length. There are four patch reefs in total separated by between 100 and 200 feet (30 to 60 meters) of sand.

Richard Whitcombe/Shutterstock ©

Stingrays, yellowhead jawfish and hogfish are often found in the sandy areas, while the patch reefs support various grouper species, margates and angelfish. The patch reefs top out at depths of around 45 feet (14 meters), and the largest has a few narrow swim-throughs, although these can be hard to find.

Route

Divers usually drop onto Paul's Reef at the southern end of the site and travel with the prevailing current, which usually runs from south to north. Exact placement will depend on the captain, but most aim to put divers about 1,000 feet (300 meters) south of Sloan's Curve.

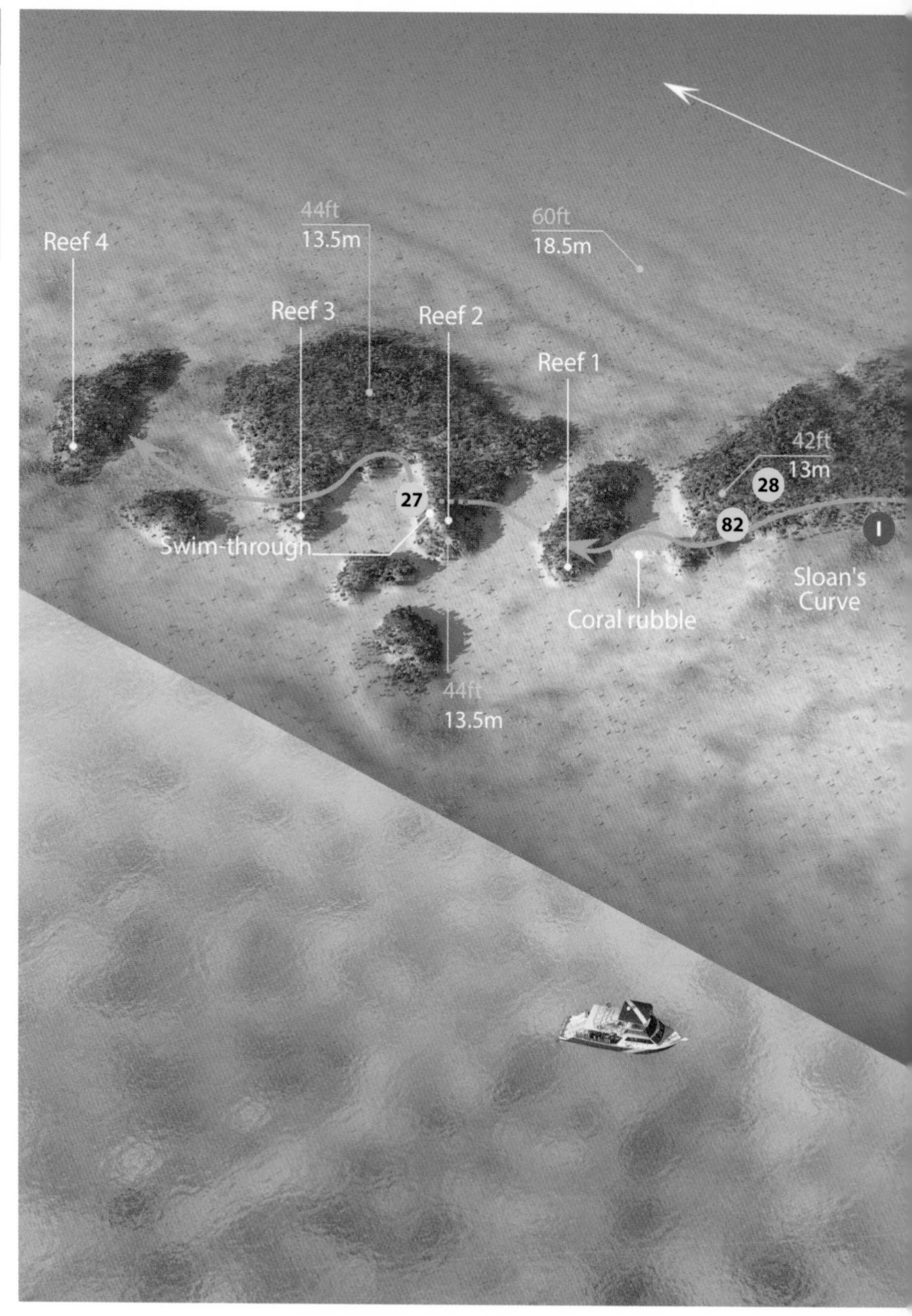

Divers can either follow the western edge of the reef or follow the eastern edge before crossing over the reef to reach Sloan's Curve. Either way, Sloan's Curve is the best starting point to swim across the sand to reach the four patch reefs that make up the northern end of this site. And to reach each of the four reefs, divers must swim in a north-northeast direction. A compass is usually required – or an experienced guide with a good sense of direction. Most divers should have enough air and bottom time to reach Sloan's Curve and then cross over to the first patch reef, but few are able to visit all four patch reefs before the end of their dive.

Other species commonly found at this site: B 1 3 4 5 17 23 31 32 35 67 68 84 85 87 90

Lofthus

Difficulty ●○○
Current ●●○
Depth ●○○
Reef ★☆☆
Fauna ★★☆

Access: about 5 mins from Boynton Inlet; 30-min walk from nearby parking; about 5 mins from shore

Level Open water

Location
Manalapan, Palm Beach County
GPS: 26°33′46.9″N, 80°02′18.6″W

Getting there

The *Lofthus* Underwater Archeological Preserve is located adjacent to the small beach side community of Manalapan, Florida. The wreck lies approximately halfway between the public beach access points at the Boynton Inlet to the south and Lantana Beach to the north.

To access the wreck from Lantana Beach in the north, head east on East Ocean Avenue across Lake Worth Lagoon to Hypoluxo Island. East Ocean Avenue ends at the Lantana Beach access, where the Lantana Municipal Beach Parking lot is located. There is ample paid parking here, at a cost of $1.50 per hour. The wreck of the *Lofthus* is located about 1.5 miles (2.4 kilometers) south along the beach – a walk of about 30 minutes.

Alternatively, the wreck can be accessed from the south by parking at Ocean Inlet Park. There is ample parking and it is free. The beach can be accessed immediately to the north of the bridge that spans the Boynton Inlet. The wreck is located about 1.25 miles (3 kilometers) north along the beach – a walk of about 25 minutes

Access

The wreck of the *Lofthus* is located approximately 550 feet (167 meters) from shore in about 6 meters (20 feet) of water – a swim of about 5 minutes. The water is relatively easy to access along this stretch of beach. Locating the wreck can be fairly challenging as she is not marked with a buoy. Most people enter the water in front of the third large house on the beach to the north of the point where South Ocean Boulevard veers slightly away from the shore.

Description

The *Lofthus* was a British-built, iron-hulled sailing ship launched in 1868. She saw service as a cargo ship during the late 19th century. The vessel ran aground on Florida's east coast on February 4, 1898, while en route to Buenos Aires with a cargo of lumber. The crew, including the ship's dog and cat, were rescued and attempts were made to pull the ship into deeper water, but she was stuck fast and could not be saved. With the ship lost, attention turned to salvaging the valuable 800,000 feet of lumber that she carried. Blasting was eventually required to give salvagers access to the cargo, which caused significant damage to the wreck.

The *Lofthus* dive and snorkel site from above.

Today, the wreck of the *Lofthus* sits broken-up and spread over an area equivalent to half the size of a football field, thanks in large part to the pounding Atlantic surf. The wreck was nominated as Florida's

eighth Underwater Archaeological Preserve in 2001 and she was listed on the National Register of Historic Places in 2003.

It is hard to grasp how magnificent a ship she once was as sections of the wreck are often hidden beneath the shifting sands of the shallow seabed. Divers and snorkelers can often make out the shape of the iron hull however, which lies roughly parallel to the beach with its bow to the north. Toward the stern, a section of one of the ship's three masts is often visible, as well as pieces of decking and timber beams that can rise as high as 6 feet (2 meters) from the bottom, depending on how the sands have shifted.

Route

The *Lofthus* can be visited by boat, or by swimming from shore. Orienting on the wreck can be quite tricky, but divers and snorkelers can remember that the bow points roughly north. The best way to explore this site is to first identify the two large deck beams that are the most obvious sections of the wreck. Divers and snorkelers can then move

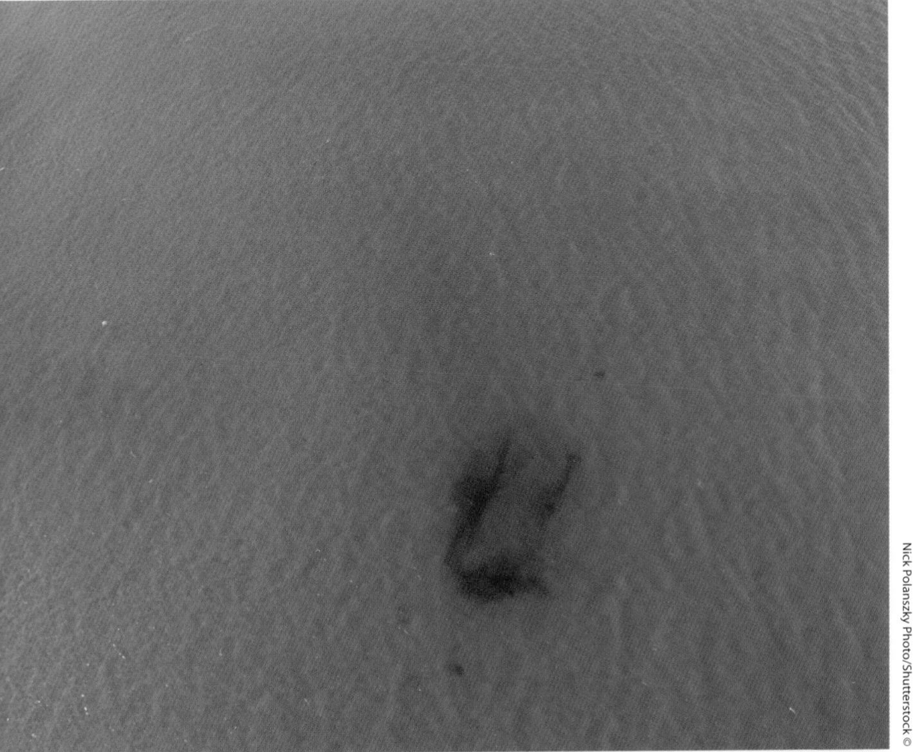

Nick Polanszky Photo/Shutterstock ©

DID YOU KNOW?

Wrecks of metal-hulled sailing ships, such as the *Lofthus*, are incredibly rare. These unique vessels were only constructed for a period of about 50 years in the latter part of the 19th century. Shipbuilders had been making wooden-hulled sailing ships for centuries, but iron-hulled, and later steel-hulled, sailing ships started to appear in the mid 1850s. These ships were constructed mainly in the U.K., and they proved stronger, cheaper, safer and lighter (and therefore faster) than their wooden counterparts. The manufacture of iron sailing ships surpassed wooden ones around 1870, putting the huge Canadian wood-building shipyards in peril. But by the time *Lofthus* sank in 1898, the market for metal-hulled sailing ships was already crashing. Major advances in steam engines made this new technology better suited for long distance shipping relative to sail power. Moreover, the insurance costs for sailing vessels were soaring. The "Age of Sail" had come to a close and the "Age of Steam" had begun.

A - the original *SV Lofthus*
B - elements of the wreck often hidden beneath the sand
C - elements of the wreck usually visible above the sand

northward to search for the bow. Nurse sharks, stingrays and barracuda are common around the site, as are schools of grunts and snapper. Those able to descend to the seabed can look for scorpionfish hiding on top of the wreck and for lobsters hiding underneath. Boats are asked to anchor in the sand near the wreck, rather than directly over the site, to avoid damaging this historic site.

Name:	*SV Lofthus*
Type:	Iron-hulled barque
Previous names:	*Cashmere*
Length:	223ft (68m)
Tonnage:	1,277grt
Construction:	Sunderland, U.K., 1868
Last owner:	Henschien & Co., Norway
Sunk:	February 4, 1898

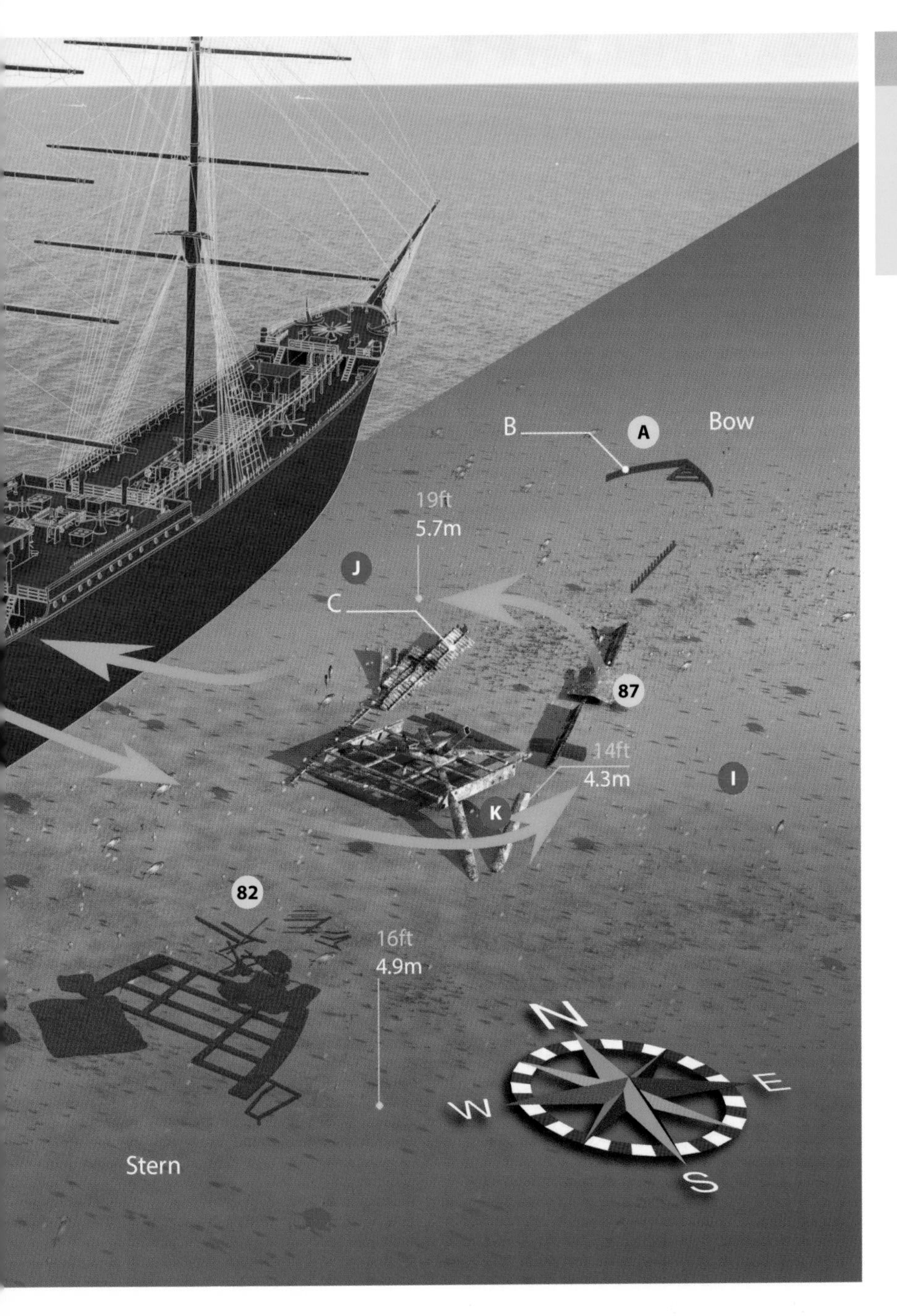

Other species commonly found at this site: 9 14 16 33 39 48 50 52 55 57 60 67 73 78 86 88

Ocean Inlet Park Reef (Boynton Inlet)

Difficulty ● ○ ○
Current ● ○ ○
Depth ● ○ ○
Reef ★☆☆
Fauna ★☆☆

Access about 14 mins from Boynton Beach
about 5 mins from shore

Level Open Water

Location

Boynton Beach, Palm Beach County
GPS: 26°32′35.4″N, 80°02′31.3″W

Getting there

Ocean Inlet Park Reef is a shore-accessible series of artificial reefs that run parallel to the beach, immediately south of the Boynton Inlet. The beach is just across the road from Ocean Inlet Park, which sits on the Intracoastal Waterway. There is plenty of free parking associated with the park, along with bathroom and shower facilities. To get there, drive east on Boynton Beach Boulevard. Turn south on Route 1 for 0.1 miles (161 meters) before heading east again on E Ocean Boulevard. Once over the Intracoastal, turn north on Hwy A1A. After 1.2 miles (2 kilometers) turn left into the park. The beach will be visible on the opposite side of the road. The official address is 6990 North Ocean Boulevard, Boynton Beach, Florida 33435.

Access

This site is accessible to snorkelers of all experience levels and may also be interesting to novice divers. A crosswalk leads from the parking lot to the beach adjacent to the lifeguard tower. Snorkelers and divers should check on ocean conditions at the lifeguard tower before entering the water. The artificial reefs are about 400 feet (120 meters) off shore from the beach, depending on the tide and the size of the beach at that time of year. As snorkelers and divers enter the water, they should pay attention to where they place their feet to avoid stepping on any reef organisms or rocks that may exist on the seabed. The use of a dive flag is mandatory at this site due to significant boat traffic.

Description

This set of artificial reefs is officially named the Ocean Ridge Mitigation Reef. It is a snorkeler- and diver-friendly artificial reef located 260 feet (80 meters) south of the Boynton Inlet. It consists of a line of rock piles that stretch nearly 0.45 miles (0.7 kilometers) from the inlet down to the beach in front of a condominium development. The county deployed 9,200 tons of rock and 160 tons of concrete modules in 2009 and 2010 to offset the impact on reef habitat of the nearby shoreline stabilization efforts. The project aims to further protect the shoreline while enhancing habitat for reef fish species.

Boynton Inlet from above.

These artificial reefs are home to a variety of juvenile reef fishes as well as adults of species often associated with back reef areas. Snorkelers and divers have the chance to see plenty of doctorfish, sergeant majors, bar jacks, and a variety of damselfish species. Wrasses, grunts and juvenile parrotfish are also commonly seen taking advantage of the complex habitat created by the rock piles. Coral cover is typically limited in shallow habitats such as this, although snorkelers and divers can still see colorful encrusting corals as well as the stubby blades of stinging fire coral. The rocks are covered in macroalgae and turf algae, which help support the many herbivorous species that call these reefs home.

The reefs sit on a sandy seafloor that bottoms out at a depth of just 14 feet (4 meters), which is why this site is so accessible to snorkelers. The tops of the rock piles and reef modules vary from 2 to 8 feet (0.5 to 2.5 meters) below the surface of the water. Visibility at the beach can be poor in high surf conditions, but generally improves the farther swimmers move from the beach.

Route

A typical route involves entering the water south of the lifeguard tower, right next to the large sand discharge pipe that rises out of the beach here. This entry point helps keep

Aaron Tindall/Shutterstock ©

RELAX & RECHARGE

Just over a mile to the south of Ocean Inlet Park is a restaurant that has been a popular culinary landmark for decades: **Banana Boat**. The restaurant originally opened its doors in Fort Lauderdale in 1971, but moved to its current location on the Intracoastal Waterway in 1978. It now boasts 250 feet (76 meters) of boat dockage courtesy of its prime waterfront location, along with ample parking.

The restaurant has live music, indoor and outdoor seating, a wide selection of cocktails, a three-hour "happy hour" and a closing time of 2am!

The menu is incredibly varied, featuring numerous fish dishes, steak, ribs, burgers and sandwiches, as well as salads, including a great seafood cobb. And do not forget to leave room for their desserts, notably their delicious banana-based specialties.
Visit: **Bananaboatboynton.com**

snorkelers and divers away from the busy Boynton Inlet boat traffic, while also reducing the risk they will overshoot the artificial reefs. Given the heavy boat traffic, it is essential that snorkelers and divers carry a dive flag while in the water. Once they have swum the 400 feet (120 meters) out to the artificial reefs, most snorkelers choose to head south to explore the bulk of the reefs. Once finished with their exploration, divers and snorkelers can exit anywhere along the beach.

OCEAN INLET PARK REEF

Other species commonly found at this site: O 1 6 8 9 15 16 19 24 32 37 46 48 68 84 90

Lynn's Reef

Difficulty ●●○
Current ●●○
Depth ●●○
Reef ★★★
Fauna ★★☆

Access about 12 mins from Boynton Inlet

Level Open water

Location

Ocean Ridge, Palm Beach County
GPS: 26°31′08.2″N, 80°01′58.4″W

Getting there

Lynn's Reef is located on the northern end of the reef tract that sits a mile (1.6 kilometers) off shore from the town of Ocean Ridge. It is only accessible by boat due to its distance from shore and the best way to reach it is through one of the local dive operators using the Boynton Inlet. The boat ride to the dive site is relatively quick given the site is just under 2 miles (3 kilometers) from the inlet.

Access

This site is accessible to divers of all experience levels. The north-south orientation of the ledge means it lies parallel to the prevailing currents. Most divers explore this site as a drift dive starting at a point along the wall to the south and ending after they clear the northern tip of the reef tract and start crossing into open sand. Visibility is generally good, and currents are typically moderate, both of which help divers explore the ledge at a reasonable pace while still making it possible to pause near any areas of interest. There are no mooring buoys present so divers should use a surface marker buoy.

Description

Lynn's Reef is one of the best quality sections of southeast Florida's reef tract accessible through the Boynton Beach Inlet. The reef is reportedly named after a local owner/operator, Captain Lynn Simmons. Other names include Vo-Ne's Reef and even Boynton Ledge, according to some charts. Lynn's Reef covers the northern section of the reef tract that runs parallel to this part of the coast. As with other ledge dives in the region, Lynn's offers divers the chance to explore a well-defined ledge in relatively shallow water or to explore spur and groove habitat on the deeper edge of the reef.

The ledge varies in height from just a few feet (1 meter) to as much as 10 feet (3 meters) in some places, and generally bottoms out on a sandy seafloor at a depth of 60 feet (18.5 meters). The ledge is easier to follow in the southern part of the reef than it is farther north, where it breaks down into a series of patch reefs and spreads toward the shore. The ledge itself disappears for long stretches before reforming farther to the north. The reef crests at a depth of 45 feet (13.5 meters).

The reef tract has a slight bend to the west near its northern terminus. Divers will need to adjust their heading and swim in a northwesterly direction to avoid getting carried over the reef crest by the current and onto the deeper spur

SCIENTIFIC INSIGHT

Sponges have existed in Earth's oceans for at least 500 million years. They are one of the oldest types of animals on our planet, and are very different from other life forms. For instance, sponge cells do not have specialized functions, meaning that each one individual cell can do any and all jobs that any other sponge cell can do. This lack of specialization also plays a role in the fact that sponges can regenerate from fragments. When violent storms break apart larger sponges, those pieces are spread across the reef where they can potentially take root and start growing again.

Another difference: Giant tube sponges can live as long as 2,000 years (maybe longer), and they can grow up to 6 feet (2 meters) across. So the next time you go diving and see a particularly large one, know that it might have been around since the time of the Ancient Romans.

and groove habitat on the east side of the reef. No matter where divers choose to explore on this reef, they will have plenty to see.

Parts of the reef patches and ledge sections to the east are heavily undercut, creating interesting habitat for a range of species, from lobsters to squirrelfish and even resting nurse sharks. The reef is covered in a mix of hard and soft corals, large barrel sponges and dense stands of gorgonians, while the space between patch reefs is a mix of rubble and sand that supports plenty of marine life. Divers can see schools of yellowtail snapper, black margates, butterflyfish, multiple species of parrotfish and angelfish, and even a few patrolling barracuda.

Once through the wide region of patch reefs, divers will encounter a second distinct ledge to the west which represents a transition from reef to open sand. The sand here is clear of rubble and it bottoms out at a depth of 65 feet (20 meters). Divers can follow this ledge as it curves around to the northeast and meets up with the eastern reef edge of the spur and groove habitat.

Lawrence Cruciana/Shutterstock ©

A great barracuda cruises above the reef.

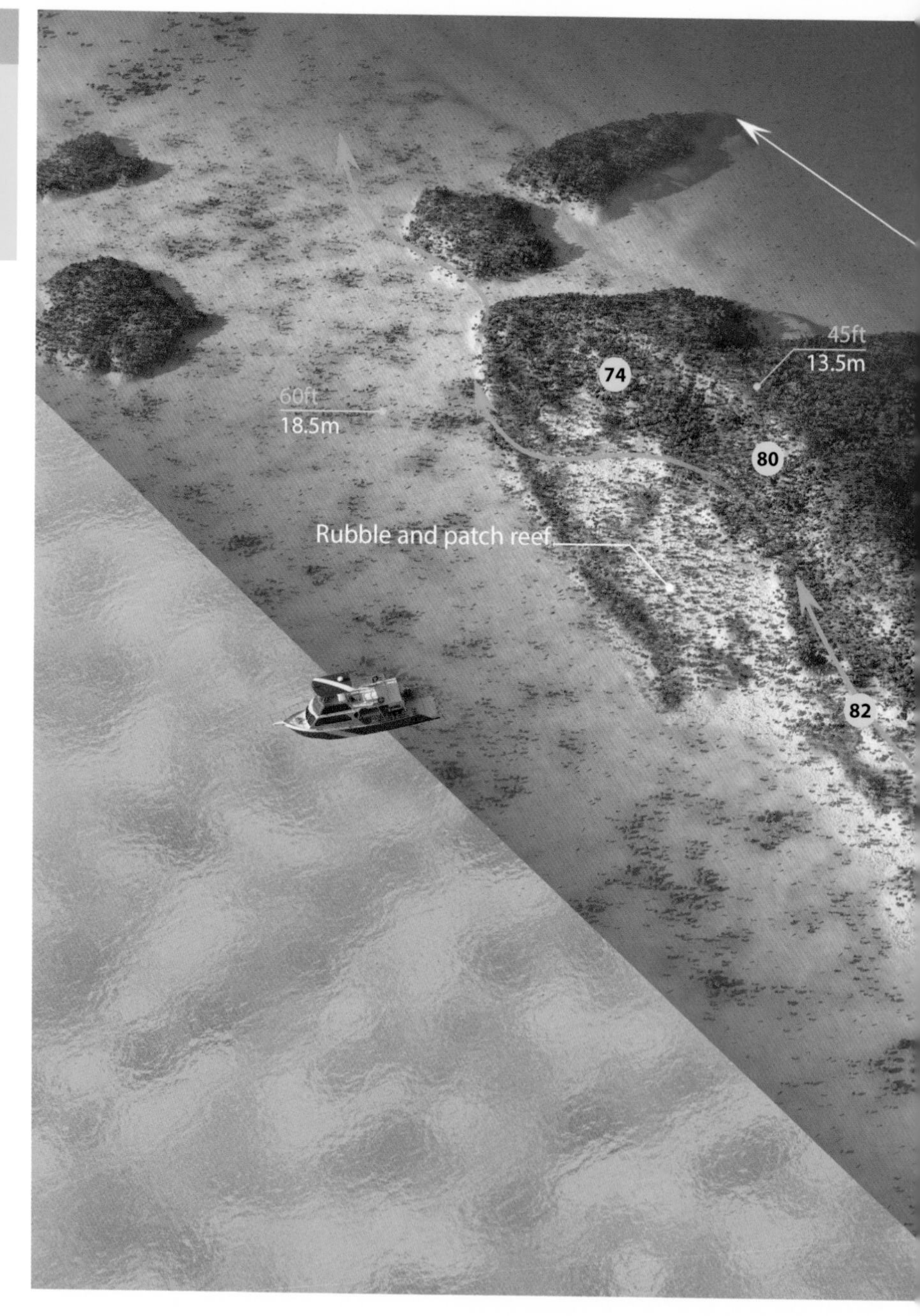

Route

A typical route for Lynn's reef involves drifting from south to north along the ledge until the contour of the reef starts becoming harder to follow. The exact location of the drop site will depend on the dive operator as well as the part of the reef divers want to explore on their dive.

Many divers choose to swim out into the wide plateau of patch reefs after losing contact with the ledge, checking out the small caves and crevices formed by the undercut patch reefs. Drifting through the patch reef offers plenty to see and explore. Divers will eventually reach the western ledge with its well-defined interface with clear sand.

If bottom time and air supply allow, divers can cut across the open sand to reach the large reef patch that sits around 200 feet (61 meters) to the north.

Other species commonly found at this site: 5 6 13 15 26 30 40 43 48 49 56 64 73 76 89 92

Boynton Ledge

Difficulty ●●○
Current ●●○
Depth ●●○
Reef ★★★
Fauna ★★☆

Access about 22 mins from Boynton Inlet

Level Open Water

Location

Gulfstream, Palm Beach County
GPS: 26°29'43.6"N, 80°02'07.8"W

Getting there

Boynton Ledge is a section of reef ledge that runs parallel to the shore 1 mile (1.6 kilometers) off shore from the coastal town of Gulfstream. It is only accessible by boat due to its distance from shore and the best way to reach it is through one of the local dive operators operating out of the Boynton Inlet. The 3.4-mile (5.5-kilometer) boat ride from the inlet takes just over 20 minutes, depending on conditions.

Access

This site is accessible to divers of all experience levels. Because of the north-south orientation of the ledge, it lies parallel to the prevailing currents. Most divers therefore explore this site as a drift dive. Visibility is generally good and currents can be strong, which helps divers explore more of the ledge on their dive. There are no mooring buoys present so divers should take a surface marker buoy with them as they dive.

Description

Boynton Ledge is a long stretch of reef that parallels the shoreline. The site is sometimes referred to by other names, including Unteidi's Reef and Tumbled Rocks. Depending on the dive operator, the term Boynton Ledge may also apply to a site about 1.5 miles (2.4 kilometers) to the north. The reef in this area is one continuous stretch of ledge and offers the opportunity for a fun drift dive.

The west side of the ledge is perhaps the most commonly explored side of the reef; it bottoms-out on sand at around 60 feet (18.5 meters) making it suitable for divers of all experience levels. The ledge itself has numerous crevices and undercuts worth exploring, particularly during lobster season.

Barrel sponges, mixed with hard and soft corals are typical of the east side of Boynton Ledge.

As with many of the reef ledges just off the coast of Palm Beach County, Boynton Ledge also offers divers the opportunity to cross over the reef and experience the deeper spur and groove formations along the eastern edge. The sand and rubble-bottomed channels of the spur and groove area reach a depth of 80 feet (24.5 meters), requiring a little more experience from divers wanting to explore here. The tops of the spurs reach up to 65 feet (20 meters). The spur and groove area rarely disappoints divers who venture into these deeper waters. The reef here is densely covered in giant barrel sponges, soft corals, and a forest of gorgonians. Devil's sea whips are sprinkled throughout the site, often extending their long, curved forms into the water column above the rubble-bottom grooves.

The complex habitat of this reef supports a variety of fish species, including blue parrotfish, stoplight parrotfish, reef butterflyfish, honeycomb cowfish, rock beauties and Queen angelfish, among many others. Divers will encounter Atlantic spadefish schooling above the reef, as well as the occasional patrolling barracuda.

Currents can be strong along this stretch of coast. While a fast current can help divers cover a lot more ground, it can also make it difficult to pause over a single spot on the reef and truly experience the ebb and flow of life there. Fortunately, the grooves are deep enough to shelter divers from

Ethan Daniels/Shutterstock ©

the current in order to observe the reef from a close vantage point.

Route

Most divers drift from south to north with the prevailing currents, whether along the ledge or across the spur and groove section. Depending on the dive operator, and even the size of the group, divers may be dropped anywhere along this stretch of reef. Divers should pay attention to their heading if they decide to head over to the east side of the reef, as it can be disorienting to cross the reef crest.

Air divers will be fine diving on the ledge but may find their bottom time limited if they choose to dive the deeper spur and groove section. Diving on nitrox can help ensure enough time to fully explore the reef here.

60ft
18.5m
C
80ft
24.5m
27
67ft
20.5m
W
S
N
E

Other species commonly found at this site: A B 1 4 9 11 13 15 19 22 26 32 35 64 65 76

MV Becks

Difficulty ●●○
Current ●●○
Depth ●●●
Reef ★★☆
Fauna ★★☆

Access about 30 mins from Boynton Inlet

Level Advanced Open Water

Location

Gulfstream, Palm Beach County
GPS: 26°28'51.7"N, 80°02'20.8"W

Getting there

MV Becks sits a mile off shore from the stretch of coastline that sits between the town of Gulfstream and the city of Delray Beach. It is nearly 4.5 miles (7 kilometers) south of the Boynton Inlet and requires a boat ride of around 30 minutes to get there. The site is best reached through one of the area's local dive operators.

Access

There is no permanent mooring buoy on the wreck, so local operators often put divers in the water slightly up current from the site. Currents typically flow north along the coast at this site although they can shift, even over the course of a dive, so be sure to keep an eye on your compass heading as you move between the sections of the wreck. *MV Becks* rests just 800 feet (240 meters) north of two other popular dive sites, *Budweiser Bar* and the *Castor*, but these sites are most commonly explored as a single-wreck dive and not as a trek. Due to its depth, the site is best accessed by experienced divers with an advanced open water certification. Diving on nitrox will help ensure divers have enough time to truly explore the site. Visibility is generally good, and penetration is not recommended.

Description

MV Becks is informally known as *Captain Tony*, after Captain Tony Townsend, a local dive charter captain who was killed by a drunken boater in the same year the artificial reef was deployed. *MV Becks* was originally a 175-foot (53.3-meter) cargo ship built in 1961 in the Hamburg-based shipyard of Deutsche Werft A.G. She was christened with the name *MV Spree* under the ownership of German shipping company, Schepers Rhein-See Linie, and operated under that name for a little over a decade. From 1974 through 1995, however, she changed ownership a dozen times, with multiple name changes during that period. Throughout the 1970s, she was owned by various British shipping companies, including one that had her hauling timber. Sold and reflagged as a Honduran vessel in the 1980s, she reportedly spent the next decade or more carrying cargo between Haiti and Miami.

Authorities seized her in 1995 when she was caught smuggling cocaine into Miami. According to news reports at the time, she was put up for auction and purchased by a new owner who was killed in a mugging before the vessel could be returned to service. The county's artificial reef program acquired her for use as an artificial reef soon after, and sent her to the bottom on October 22, 1996, as the program's 29th artificial reef.

Becks currently sits in two large pieces: The bow to the south and the stern to the north. A large debris field, stretching nearly 115 feet (35 meters), lies directly south of the stern. The bow section sits tilted heavily to port with its mast pointing out at a northeastern heading, just slightly above horizontal. The stern rests on its starboard side, and its front section lies mangled where it was torn from the main cargo holds, which also lie in ruins. The average depth of the site is 90 feet (27.5 meters) on what is primarily a sandy bottom.

This wreck is very popular with goliath grouper during the summer months –

SCIENTIFIC INSIGHT

During the 1990s, coastal freighters were popular targets for artificial reef programs. They were readily available, thanks to the drug war of the '80s and '90s that led authorities to seize boats caught attempting to smuggle drugs into the U.S. The thick steel hulls and large open holds of these freighters provided

Leonardo Gonzalez/Shutterstock ©

Schooling porkfish are a common sight throughout Palm Beach County.

habitat for reef fish and plenty of areas for divers to explore.

The underwater environment is unforgiving, however, and the open cargo holds of these vessels often collapsed into debris fields, which can damage the fragile marine life growing on the structure. Once the cargo areas collapse, the bow and stern sections often become separated – in rare cases ending up miles apart.

For these reasons, artificial reef programs refocused their efforts in securing more robust vessels to use as artificial reefs. The *Ana Cecilia* is an excellent example of a sturdier vessel that will hopefully provide decades of complex habitat for marine life and divers.

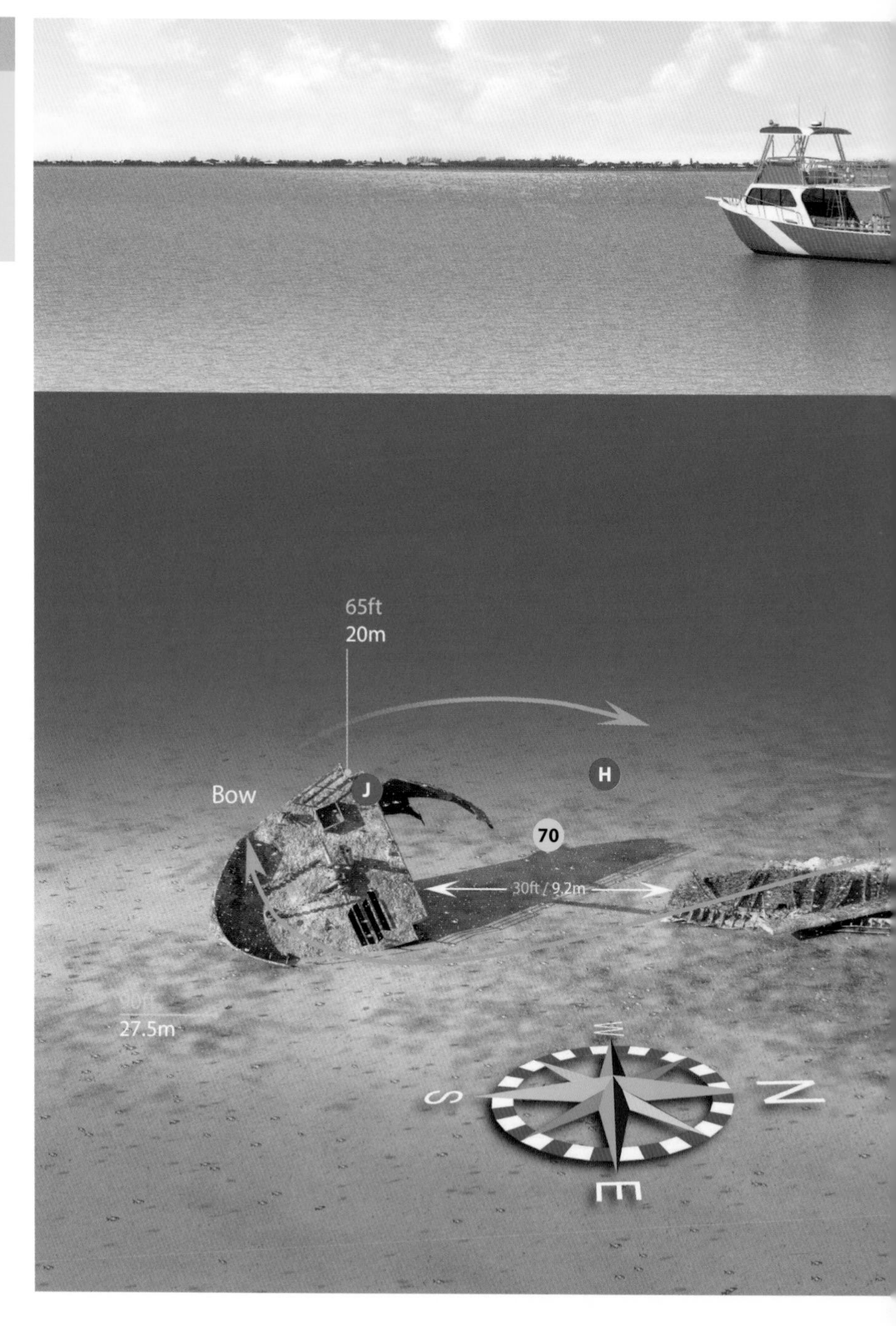

divers may see them clustered around the length of mast that sticks out of the bow section. Porkfish, black margates and grey snapper are often seen around the two intact sections and among the pieces of debris that are scattered throughout the site, along with grunts, goatfish and various species of parrotfish. The hull, superstructure, railings and pretty much every other surface of the stern and bow sections are heavily encrusted with hard and soft corals, courtesy of nearly three decades spent underwater.

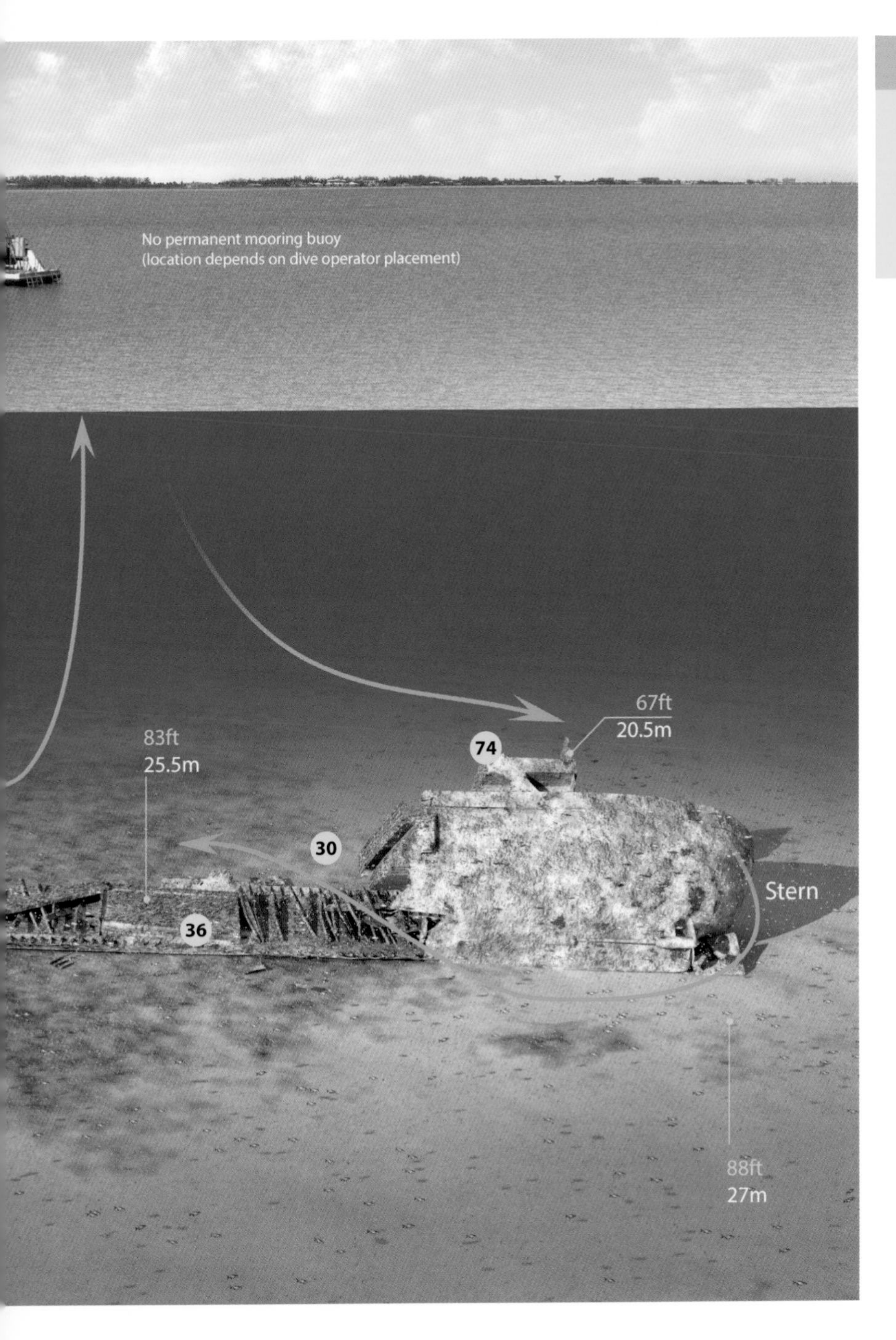

Other species commonly found at this site: 2 5 7 8 11 13 16 20 22 25 26 31 49 66 68 71

Route

The route will depend on the currents at the time of the dive. In a strong current, it may not be possible for divers to move easily from one section of the wreck to the other against the current. In low-current conditions, divers will have ample time to circle both the bow and stern sections and back again. Divers can take extra time to check out the small spaces under the central hull sections and the supporting cross braces that lie scattered in the debris field. During the grouper breeding season, divers can float near the bow's mast and watch the goliath grouper congregate nearby.

Name:	*MV Becks*	**Length:**	175ft (53.5m)
Type:	Coastal freighter	**Tonnage:**	425grt
Previous names:	*MV Christ Capable, MV Paul-Marie, MV Sabine, MV Kimble, MV Barcourt, MV Harcourt, MV Timber Queen, MV Spree*	**Construction:**	Deutsche Werft A. G., Hamburg, 1961
		Last owner:	Unknown
		Sunk:	October 22, 1986

65ft
20m
J
70
30ft / 9.2m
H
Bow
90ft
27.5m
30
S
W
E
N

No permanent mooring buoy
(location depends on dive operator placement)
83ft
25.5m
67ft
20.5m
36
74
Stern
88ft
27m

Budweiser Bar

Difficulty ●●○
Current ●●○
Depth ●●○
Reef ★★☆
Fauna ★★☆

Access about 32 mins from Boynton Inlet

Level Advanced Open Water

Location

Gulfstream, Palm Beach County
GPS: 26°28′43.5″N, 80°02′18.8″W

Getting there

Budweiser Bar rests a mile off shore from the stretch of coastline between the town of Gulfstream and the city of Delray Beach. It is nearly 4.5 miles (7 kilometers) south of the Boynton Inlet – around a 30-minute boat ride. The site is best reached through one of the area's local dive operators via the Boynton Inlet.

Access

There is no permanent mooring buoy on the wreck, so local operators will often put divers in the water slightly up current from the site. Currents typically flow north along the coast at this site although they can shift, even over the course of a dive, so be sure to keep an eye on your compass heading when moving between the different sections of the wreck. *Budweiser Bar* rests just 800 feet (240 meters) south of another popular dive site, *MV Becks* and just west of the *Castor*.

Each of these sites is most commonly explored as a single-wreck dive and not as a trek despite their relative proximity. Visibility is generally good here, although it can be highly variable. Due to its depth, the site is best accessed by experienced divers with advanced open water certifications. Diving on nitrox will help maximize bottom time and allow divers to truly explore the site. Penetration is not recommended on this wreck.

Description

Budweiser Bar was sunk as an artificial reef in July 1987. Built in Germany in 1965, the 167-foot (51.5-meter) freighter plied the Caribbean transporting cargo between the U.S., Haiti and the Bahamas. She was originally christened the *MS Havel* and was owned and operated by Schepers Rhein-See Linie. In 1977, the German shipping company sold her, and her new owners

A diver explores a wreck.

renamed her *Olive M.* and operated her under a Dutch flag. Her name changed one last time just before the county scuttled her as *Budweiser Bar,* after the Budweiser Bar-Brown Distributing company (the West Palm Beach distributor of Anheuser-Busch products at the time) raised the necessary money to deploy the ship as an artificial reef.

The wreck rests on a sandy seafloor at a depth of 97 feet (29.5 meters), but the decades spent

RELAX & RECHARGE

Hurricane Alley is located just south of the Boynton Inlet – the preferred departure spot for trips to *Budweiser Bar* and the adjacent wrecks. This popular raw bar and restaurant has been "blowing away" its customers for years. Hurricane Alley touts its oversized portions, so you and your appetite will not be disappointed. Their Lobster Mac 'n' Cheese is a menu favorite, along with the soft-shell crab, a half-pound cheeseburger and prime ribs offered on Friday and Saturday nights. It is a relaxed joint where even dogs are catered to – they can order from the "doggie bites" menu with the help of their owners. If you only want a snack and a drink, then stop in during happy hour, which runs from 3pm to 6pm on weekdays and features clams or oyster specials. You can even upsize your drink to a 32oz "storm bucket."
Visit: **Myhurricanealley.com**

Peter Leahy/Shutterstock ©

underwater have not been kind. The relatively intact stern section lies on its starboard side, slightly to the north of the mostly flattened bow section. The prow of the bow section sits mostly upright, almost 15 feet (5 meters) above the seafloor. The bow section itself is close to 117 feet (35.5 meters) long and represents the central cargo hold area of the coastal freighter. Despite the multiple cross braces installed to support the hull, the cargo hold lies mostly flat.

While not much is left in terms of penetration opportunities on the *Budweiser Bar,* it remains an interesting dive. The years underwater have allowed hard corals, gorgonians and sponges to heavily colonize the surface of the wreck. Both the bow and stern sections are teeming with marine life. Divers will have plenty of opportunity to see barracuda, Spanish hogfish, numerous species of grunt, sergeant majors and the ubiquitous schools of porkfish. During the summer months, this wreck is also popular with goliath grouper.

Route

The route depends on the currents at the time of the dive. In a strong current, it may not be possible for divers to move easily from one section of the wreck to the other against the current. Divers can shelter behind the bow section to explore the remains of the cargo hold, while the stern section also provides ample shelter from strong currents. Under low-current conditions, divers will be able to circle both the bow and stern sections in a figure eight pattern.

Other species commonly found at this site: J 2 3 6 7 8 11 15 21 22 26 35 37 57 65 91

95ft
29m
70
C
31
83ft
25.5m
97ft
29.5m
Bow

Name:	*Budweiser Bar*	**Last owner:**	Unknown
Type:	Cargo freighter	**Sunk:**	July 16, 1987
Previous names:	*Olive M., MS Havel*		
Length:	167ft (51.5m)		
Tonnage:	384grt		
Construction:	Deutsche Industrie Werft A.G., 1965		

MV Castor

Difficulty ● ● ○
Current ● ● ○
Depth ● ● ●
Reef ★☆☆
Fauna ★★★

Access about 32 mins from Boynton Inlet

Level Advanced Open Water

Location

Delray Beach, Palm Beach County
GPS: 26°28′43.8″N, 80°02′14.3″W

Getting there

The wreck of the *MV Castor* lies 4.6 miles (7.5 kilometers) south of the Boynton Inlet and about 1 mile (1.7 kilometers) off shore of Delray Beach. The site is only accessible by boat due to its distance from shore and the best way to reach it is through one of the local dive operators using the Boynton Inlet. The travel time from the inlet is about 30 minutes, depending on conditions.

Access

There is no permanent mooring buoy on the *MV Castor*, but as the current can be strong at this site, most dive operators attach a line to either the bow or stern so that divers can descend and

ascend more easily. The seabed is at a depth of 105 feet (32 meters), but most divers maintain a shallow depth profile as they explore the wreck.

The wreck is more suitable to advanced divers because of the depth and potential for strong currents. Visibility is generally good and divers using nitrox should have sufficient bottom time to adequately explore the site. However, those divers using air may run out of bottom time before fully exploring both sections of the wreck.

Description

MV Castor is one of the most popular wreck dives in Palm Beach County because each dive is often unique with respect to both the conditions and the species that divers encounter. The wreck is large and relatively deep and currents can change daily, sometimes even within the space of a single dive. As such, divers who visit the wreck have come to expect the unexpected.

Castor is known as a good place to see goliath grouper.

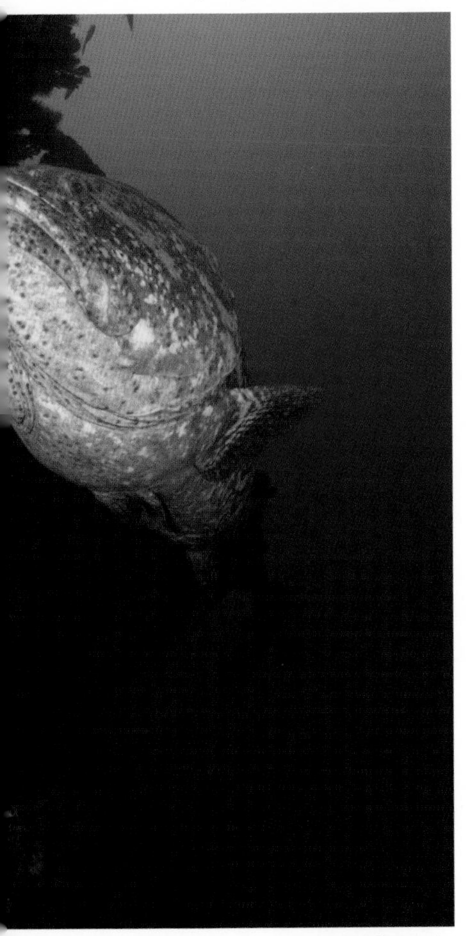

Matt9122/Shutterstock ©

DID YOU KNOW?

The seizure of *MV Castor*, and the arrest of her captain and crew on drug trafficking charges, was just the tip of the iceberg. Over the following year, more than 40 additional arrests were made in numerous different countries. MV Castor was just one of 10 ships that formed a smuggling armada operated by a Columbian drug lord and his investors. Using high-powered cigarette boats, the smugglers would ferry the drugs from their jungle manufacturing stations out to the ships, which were waiting about 100 miles (160 kilometers) off the Colombian and Venezuelan coasts. In total, authorities believe that cocaine with a street value of around $3 billion was smuggled into Europe and North America by the cartel, a third of which was intercepted by law enforcement on *MV Castor* and several of her sister ships.

The wreck is known for being a great spot to encounter goliath grouper. About a dozen individuals are known to inhabit the wreck year-round, but the wreck is also a popular spawning site. In late summer, divers can easily find themselves surrounded by a hundred of these gentle giants.

MV Castor was a coastal freighter built in the Netherlands in 1970. She operated in numerous locations and under a staggering number of names for close to three decades before she was seized by the U.S. Coast Guard off the coast of Venezuela. She had more than 10,000 pounds of cocaine on board. The ship was deployed as an artificial reef on December 14, 2001.

MV Castor originally sat upright on the seabed with her deck level at around 90 feet (27.5 feet) and her superstructure rising to within 60 feet (18 meters) of the surface. Over the years, however, the midsection of the wreck has collapsed onto the seabed, along with the hull plates, which are spread like wings to each side of the deck. The bow and stern sections separated from the rest of the structure and have now rolled onto their sides, separated by about 150 feet (46 meters) with the bow lying roughly to the north of the stern.

The shallowest part of the stern section is now about 70 feet (21 meters) and the shallowest part of the bow is about 80 feet (24 meters). Much of the structure is now covered in coral and sponge growth, thanks to the length of time spent underwater. However, the partial collapse of the wreck has limited the penetration opportunities.

Route

The entire wreck is spread over an area approximately the size of a football field with most divers looping around both the stern and bow sections of the wreck. Most dive operators tie off on the wreck's bow or stern, marking the beginning and ending points of the dive. These

Other species commonly found at this site:  H  I  K  L M 2 7 11 14 24 33 69 72 4 79 81

two sections are the most interesting to explore as they are the most complex parts of the wreck. Even so, the deck and hull plates lying on the sand between the two sections are worth exploring, as they often provide shelter for lobsters, snapper, black grouper and the occasional nurse shark.

Goliath grouper can be found throughout the wreck. However, they are most common around the larger stern section where there are more places to hide within the structure. Barracuda are frequently seen hovering in the water column throughout the site, while a large school of horse-eye jacks is often present on the bow.

No permanent mooring buoy
(location depends on dive operator placement)

103ft
31.5m

82

77

79ft
24m

110ft
33.5m

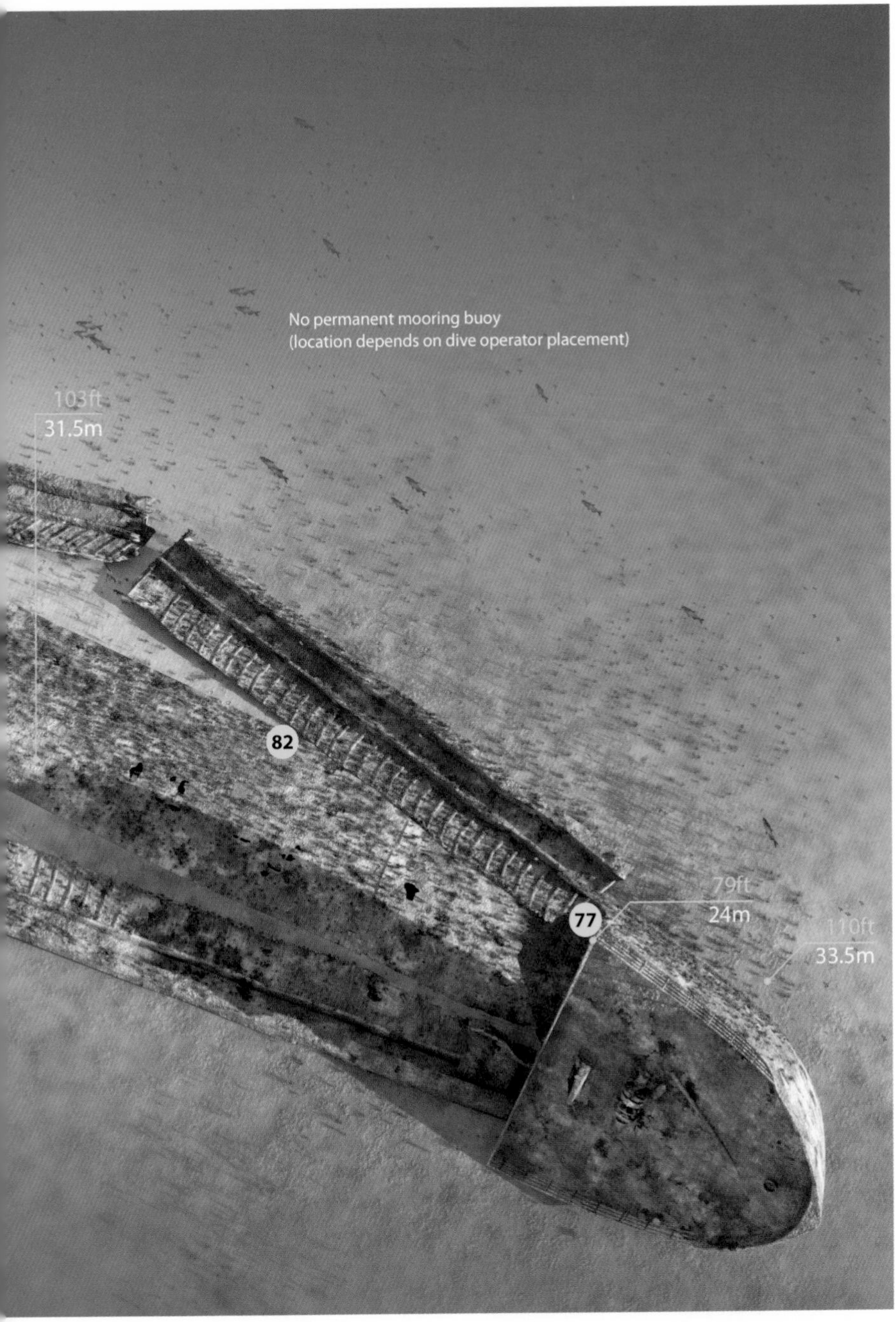

Name:	*MV Castor*	**Length:**	258ft (78.6m)
Type:	Coastal freighter	**Tonnage:**	499grt
Previous names:	*MV Dynacontainer, MV Tropic*	**Construction:**	Martenshoek, Netherlands
	MV Dorothee Bos, MV Tor Timber	**Last owner:**	Columbian drug lord
	MV Bell Chieftain, MV Irma	**Sunk:**	December 14, 2001
	MV Mer Star, MV Ann Mary, etc		

Delray Ledge

Access about 35 mins from Boynton Inlet

Level Open Water

Location

Delray Beach, Palm Beach County
GPS: 26°28'15.0"N, 80°02'35.8"W

Getting there

Delray Ledge Reef is located on a stretch of stand-alone reef ledge that runs parallel to shore just 0.75 miles (1.2 kilometers) off shore of Delray Beach. It is only accessible by boat due to its distance from shore and the best way to reach it is through one of the local dive operators that use the Boynton Inlet. The boat ride from the inlet to the dive site is nearly 5 miles (8 kilometers) and takes around 35 minutes.

Access

This site is accessible to divers of all experience levels. The north-south orientation of the ledge means it lies parallel to the prevailing currents. Most divers therefore explore this site as a drift dive starting at a point along the wall to the south and ending at the distinct hook-like northern terminus of the reef. Visibility is generally good, and currents are typically moderate, both of which help divers explore the ledge at a reasonable pace while still making it possible to pause and explore areas of interest. There are no mooring buoys present along this ledge so divers should bring a surface marker buoy with them.

Schooling grunts on a reef in Florida.

Description

Delray Ledge Reef is a mile-long stretch of reef. Some maps refer to it as Hobie Ledge. This site covers the northern section of the reef that runs parallel to the coast just off shore from Delray Beach and provides divers with two distinct habitats to explore. The primary focus of this site is the distinct, 10 to 12-foot-tall (3 to 3.5-meter) ledge that runs the length of the west side of the reef, facing the shore. The ledge bottoms out on a sandy seafloor at a depth of 68 feet (20.5 meters) and is undercut in many places, creating complex habitat with plenty of small caves and overhangs that support a variety of reef creatures. To the east, facing the open ocean, divers can explore a series of deeper, more complex spur and groove formations. This alternate area is best explored by divers with more experience given the slightly greater depth and potential need for navigation through the spurs and grooves.

The ledge feature on the western side of Delray Ledge Reef makes it easy for open water divers of all experience levels to explore. Divers can drift along the ledge, watching for a variety of reef fish, including ocean triggerfish, schoolmaster snapper and nurse sharks. Yellowhead jawfish frequent the sand and rubble at the base of the ledge while schools of creole wrasse swim in the water column immediately above the reef.

A few sloped areas and small patch reefs interrupt the otherwise continuous ledge. But divers will have no problem following the ledge until it reaches a narrow valley where the single ledge transitions into a double ledge. The top ledge continues at a northern heading, while the lower ledge curves around and heads west. This is the beginning of the large hook formation at the northern end of the reef. The main ledge and tip of the hook are separated by as much as 200 feet (61 meters) at their widest point. The inside of the hook shelters a raised plateau that bottoms out at around 60 feet (18.5 meters, and shelters soft corals, gorgonians and small coral heads.

Peter Leahy/Shutterstock ©

The leading edge of the hook consists of a raised section of reef, with ridges that run from east to west. In general, the crest of each ridge reaches a depth of 50 feet (15 meters) while the corresponding troughs are 53 feet (16 meters) deep. Seafans, barrel sponges and soft corals dominate the crests while the troughs are marked by narrow strips of sand and rubble that join up with the sandy plateau in the center of the hook. The ridges slope down to the sand in the west, which stretches off toward shore. This sharp transition between reef and sand continues all the way around the northern end of the hook to where the reef ends. A sand-bottomed channel runs north-south, feeding back into the central area of the hook, and marks the transition to the spur and groove formations that are the main features of the eastern side of the reef.

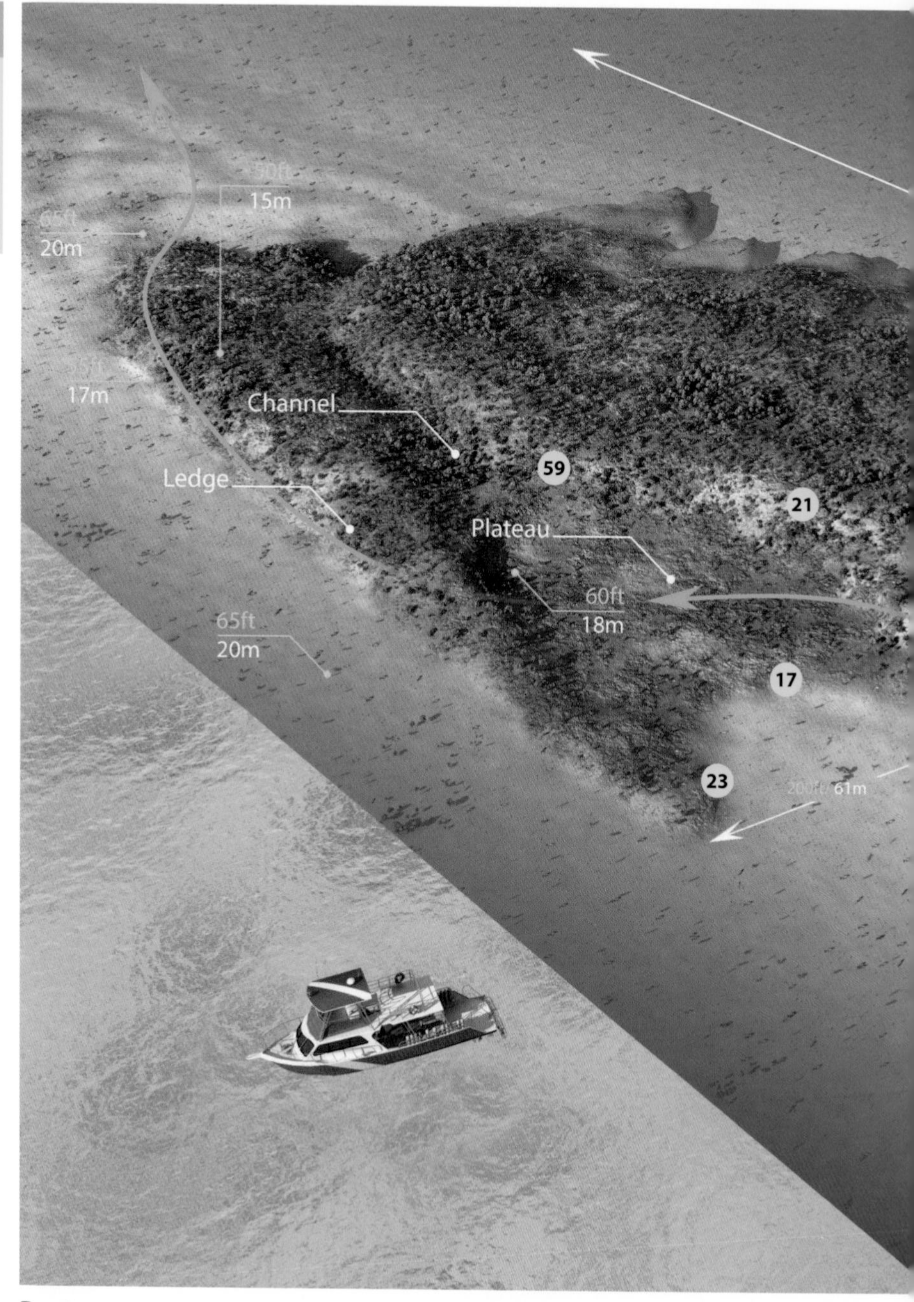

Route

Most divers drift from south to north along the ledge with the prevailing currents. The exact location of the drop site will depend on the dive operator as well as the strength of the current. Divers who wish to explore the deeper spur and groove area can swim up and over the reef to the east before continuing to drift to the north. However, most divers choose to drift along the ledge for their first exploration of this site. The ledge is easy to follow, and divers face the choice of following the top ledge to the north when the hook formation begins or following the lower ledge out toward the leading edge of the hook. Either option offers plenty to explore and see. With a variety of route options, divers often return to this site time and again.

Other species commonly found at this site: 1 8 11 12 15 26 27 31 32 34 35 38 49 61 64 75

Red Reef Park

Difficulty ● ○ ○
Current ● ○ ○
Depth ● ○ ○
Reef ★☆☆
Fauna ★☆☆

Access about 15 mins from Boca Raton
about 1 min from shore

Level n/a

Location

Boca Raton, Palm Beach County
GPS: 26°21′44.3″N, 80° 4′5.5″W

Getting there

Red Reef Park is a shore-accessible set of artificial reefs just off the beach associated with the municipal park of the same name. To get there, drive east on East Palmetto Park Road and turn north on Hwy A1A. Drive for 0.81 miles (1.3 kilometers) and turn right into the park. (If you pass the Gumbo Limbo Nature Center on your left you have driven too far north on A1A.) The park's official address is 1400 North Ocean Boulevard, Boca Raton, Florida 33432.

There is ample paid parking associated with the park on the ocean side of A1A, while a secondary parking lot across the street offers pay-by-meter parking. The park includes bathroom and shower facilities and easy access to nearly 0.5 miles (0.8 kilometers) of beach with six lifeguard towers.

Access

This site is accessible to snorkelers of all experience levels. The beach is a short walk from the parking lot and a slightly longer walk from the secondary lot on the far side of the highway. Easy-to-navigate wooden steps and a boardwalk cut through the lush shoreline vegetation to access the sandy beach. The lifeguards are on duty during normal hours (9am to 5pm). Be sure to check in at the lifeguard tower to get a sense of what conditions are like when you get there. The artificial reefs are in two small clusters near the southern end of the beach, just south of a pile of boulders that breaks the water along the shore, directly out from lifeguard tower #8. The reefs are best accessed just before high tide and when the wave break is small. The visibility can deteriorate on an outgoing tide as the inlet, located just to the south, funnels water from the Intracoastal Waterway into the ocean.

Red Reef Park as seen from above.

Snorkelers can enter the water wherever there is sand, holding their fins in their hands. The reefs are located about 150 feet (58 meters) from the beach. Snorkelers do not need a dive flag at this site.

Description

Red Reef Park Reef includes over 800 tons of rock deposited in a series of piles on either side of a shallow natural reef. There is a stretch of four artificial reefs consisting of square piles of

RELAX & RECHARGE 

The Tin Muffin Café is located on Palmetto Park Rd – a five-minute drive from Red Reef Park. It is a great lunch spot often described as one of Boca Raton's "hidden treasures." They serve a range of homemade salads, sandwiches and soups that are simply delicious, while offering patrons a cozy ambiance that may remind you of your grandmother's kitchen. There are vegetarian, vegan and even gluten-free options, and the homemade drinks include herbal raspberry iced-tea and minted lemonade. Leave room for dessert. After all, the restaurant is named after its muffins! One strange, but worthy piece of advice: Check out the bathrooms before you leave!

Reef Smart ©

rocks in an arc that runs parallel to the shore followed by another two reefs farther south, at the end of the park boundary. The total distance between the northern and southern ends of the site is around 600 feet (198 meters), and as the number of boulders is relatively small and many have been covered by sand, it can be challenging for snorkelers to locate the reefs.

The beach is also relatively exposed here, meaning the region regularly experiences waves that crash against the reefs. The wave action and shallow depth make this habitat suitable for only the hardiest of reef organisms. Snorkelers will want to be careful to avoid touching the fire corals that are common in this habitat. See the tips in the Dangerous Species section at the end of the book for what to do if you come into contact with fire coral.

The reefs also shelter juveniles of many different species, acting as a nursery for the

more developed reefs that sit farther off shore. Snorkelers have the chance to see plenty of fish swimming around the reef and the seafloor, including sergeant majors, doctorfish, bluehead wrasses, grey snapper and a variety of grunt species. Chub are common here while the occasional barracuda may be seen visiting the site.

Route

A typical route involves entering the water near lifeguard tower #8 and circling the pile of boulders that stick out of the waterline there. Snorkelers should be careful to leave enough space between themselves and the reef to avoid having the waves push them against the reef. The route continues south past the natural reef to the first set of four

artificial reefs, which are lined up, north to south, 230 feet (70 meters) south of the boulder pile. After exploring that section, snorkelers can choose whether to continue farther south to explore the remaining two reefs, which are another 230 feet (70 meters).

Other species commonly found at this site: 5 8 14 16 20 22 24 43 46 49 58 62 64 68 76 86

South Boca Inlet Reef

Access about 15 mins from Boca Raton
about 5 mins from shore

Level Open Water

Location

Boca Raton, Palm Beach County
GPS: 26°19'59.6"N, 80°04'15.0"W

Getting there

South Boca Inlet Reef is a shore-accessible series of artificial reefs that run along the beach immediately south of the Boca Raton Inlet. The beach is next to South Inlet Park. There is plenty of parking associated with this small park, along with bathroom and shower facilities. To get there, drive east on East Palmetto Park Rd over the bridge to the barrier island and Hwy A1A. Turn south and drive 1.2 miles (1.9 kilometers). Once you cross the bridge over the inlet, continue along the road for 0.3 miles (0.5 kilometers) before turning left into the park, adjacent to De Soto Road. The park's address is 1100 South Ocean Boulevard, Boca Raton, Florida 33432.

Access

This site is accessible to snorkelers of all experience levels and may be interesting to novice divers. The beach is a short walk from the parking lot through the trees and along a sandy path down to the beach, which has a single lifeguard on duty during normal hours (9:15am to 4:45pm). Be sure to check-in at the lifeguard tower to learn about the conditions before entering the water. The artificial reefs are about 400 feet (120 meters) off shore from the beach, depending on the tide and the size of the beach at that time of year. As divers and snorkelers enter the water, they should pay attention to where they place their feet to avoid stepping on the reef or any reef organisms. The use of a dive flag is mandatory at this site.

Description

The South Boca Inlet Reef consists of more than two dozen piles of rocks adjacent to two natural reefs. The county's artificial reef program placed up to 17,000 tons of rock here in 2003 to form two separate reefs – the first in close proximity to the inlet and the second 0.75 miles (1.2 kilometers) to the south. The first artificial reef stretches 700 feet (213 meters) from north to south starting at the inlet, while the second measures 600 feet (183 meters) in length.

These reefs are home to a variety of juvenile reef fish species. Snorkelers have the chance to see sergeant majors, grunts, damselfish, small wrasses and juvenile parrotfish. Chub patrol the water column above while stingrays are often spotted in the open areas of sand. Those snorkelers able to look closely enough may spot a camouflaged octopus, while juvenile green turtles have been known to cruise through the site from time to time.

The reefs are at a depth of just 12 feet (3.5 meters), which is why this site is so accessible to snorkelers and less suitable for divers. The shallowness of the site also makes visibility hit or miss. Visibility at the beach can be poor in high surf conditions, but it does often improve the farther away from shore one gets. The shallow depth means the variety and density of corals is lower than what is found on offshore reefs. But there are still colorful encrusting corals, along

An aerial shot of Boca Raton Inlet.

with fire corals and macroalgae covering the boulders, all of which contribute to the overall biodiversity of these artificial reefs.

Route

A typical route involves entering the water directly in front of the lifeguard tower. Nearly 100 feet (30 meters) off the beach, snorkelers and divers will encounter the northern section of natural reef. Turning south, they can swim 300 feet (91 meters) to cross the sand toward the artificial reefs. After exploring the reefs, head toward shore, exploring

the other natural reef in this area before getting out of the water and returning to the entry point along the shore.

SOUTH BOCA INLET REEF

Other species commonly found at this site: O 5 10 16 18 22 24 26 39 41 45 46 53 78 85 86

110 Hydro Atlantic

Access 5 mins from Boca Raton Inlet
20 mins from Hillsboro Inlet

Level Technical

Location

Boca Raton, Palm Beach County
GPS: 26°19'29.4"N, 80°03'02.6"W

Getting there

Hydro Atlantic sits near the border between Palm Beach and Broward counties, just a mile or so off shore. It is technically in Palm Beach County waters, but most dive operators who visit the wreck do so out of Broward County. Getting to the wreck requires a boat ride of just a few minutes out of the Boca Raton Inlet, or a 20-minute ride north of the Hillsboro Inlet. The best way to reach the wreck is through one of the local dive operators who use either inlet.

Access

There is no permanent mooring buoy on *Hydro Atlantic*, and placement will depend on the local operator. Visibility is generally good, but currents can be strong. The wreck sits in very deep water, at nearly 172 feet (52.5 meters), making it suitable only for technical divers. There are multiple opportunities for penetration along with many high-quality, coral-encrusted swim-throughs. *Hydro Atlantic* sank unexpectedly and was not prepared for divers. This means most of its rigging, pipes and cranes remain in place and pose a threat of entanglement. However, those same qualities also make this one of the more spectacular dives in Florida.

Description

Hydro Atlantic started her career under the name *Delaware*, as a dredger transport ship originally built for the U.S. Army Corps of Engineers in 1905. She was sold in 1950 to Construction Aggregates Corp. and renamed the *SS Sand Captain*. In 1961, she underwent a massive overhaul, swapped her steam engines for a diesel electric plant, and added an extra 1,231grt to her tonnage. She was renamed the *MV Ezra Sensibar* and remained under the same ownership. She was sold again in 1968 to Hydromar Corporation and renamed *MV Hydro Atlantic*. She sank in December 1987 while she was being towed to a salvage yard in Texas. The winds from a

The heavily colonized deck of the *Hydro Atlantic*.

DID YOU KNOW?

Technical diving differs from recreational diving in a number of key areas, but the term is generally used for any diving that exceeds the normal limits imposed on depth and bottom time for recreational divers. In short, tec divers can dive deeper and stay longer than rec divers because they have undergone the necessary training and have planned accordingly. They also use special gas mixtures that help protect them from the risks that both nitrogen and oxygen pose to divers at depth (rather than simply compressed air or even the enriched air nitrox accessible to open water divers). Furthermore, tec divers meticulously plan their dive to include sufficient decompression stops. While this means they can spend more time exploring a wreck such as the *Hydro Atlantic*, it also means they must spend more time floating at their deco stops, staring into the deep blue sea.

Matt9122/Shutterstock ©

winter storm generated waves that were too strong for *Hydro Atlantic*'s weakened hull, and she started taking on water. The tug crew were quick to release the tow ropes once they saw her fate was sealed.

While the wreck would be more accessible to divers if it had sunk closer to shore, its location and depth make it one of the top 10 wreck dives in the U.S., and one of the most well-known technical dives in Florida. She sits upright, with her deck at 145 feet (44 meters) and one of her masts reaching as high as 120 feet (26.5 meters). Her deck is criss-crossed

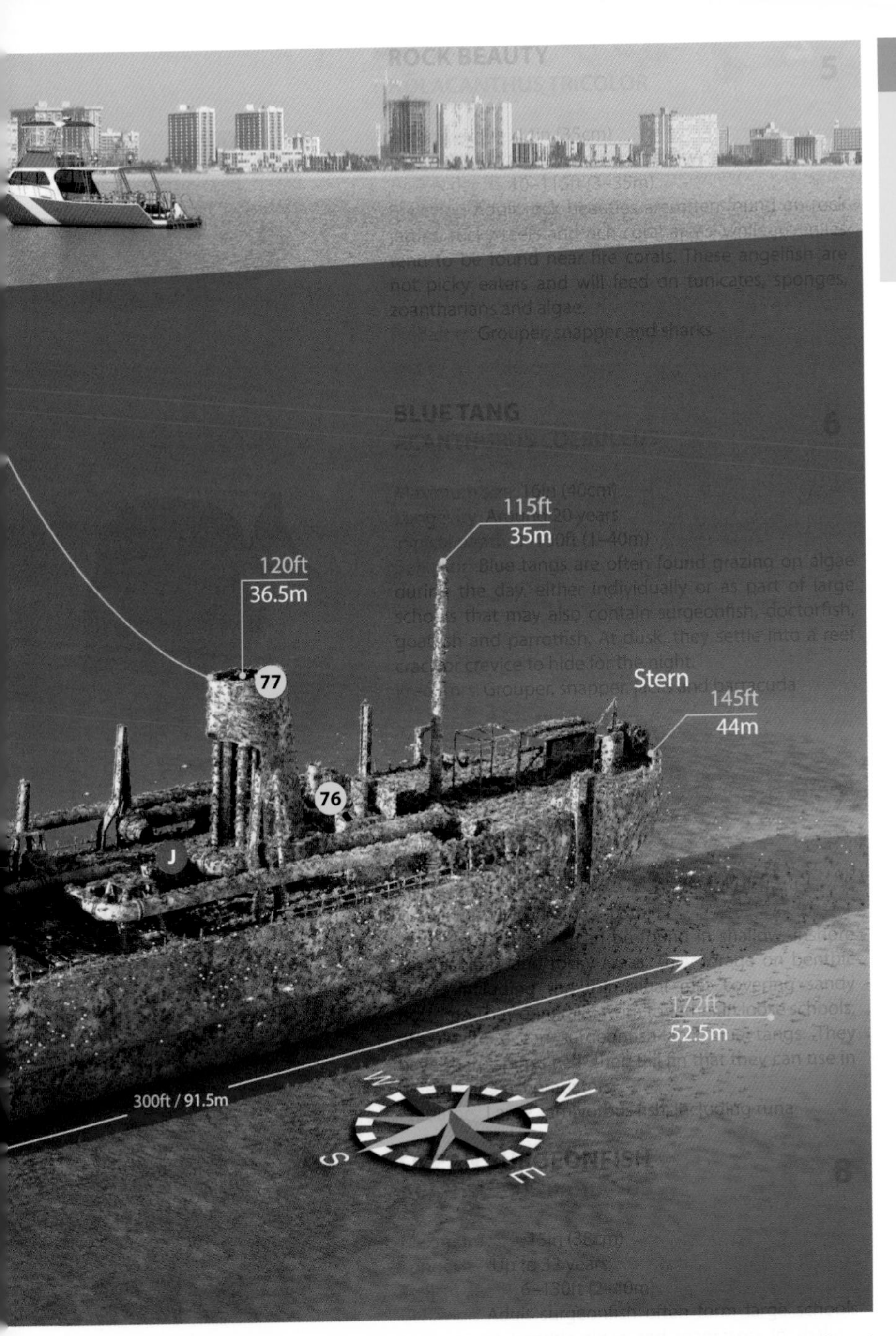

with fishing line, cables and other gear, and her original piping still runs up and down the main deck.

The wreck remains largely intact, which offers plenty of surface area for the marine life that has successfully colonized this artificial reef. Soft corals and sponges encrust the hull and the many swim-throughs, while schools of jacks and bait fish swarm the deck. Large pelagics such as barracuda are regular visitors to the wreck, and even the occasional bull shark can be spotted cruising these waters.

Route

In a fast current, divers may be dropped off well upstream of the wreck, letting them descend directly onto the wreck itself. In a slack current, however, lead divers will set a mooring line on the wreck, often on the railing surrounding the pilot house. Even as a decompression dive, there is not a lot of bottom time available for a wreck of this size, with so much to see. There are multiple penetration opportunities, with more opening up as the superstructure begins to age in the corrosive saltwater.

Divers can take their pick, either visiting the

bow or the stern first. Both options provide ample chance to explore the piping amidships as well as the penetration opportunities in the main deck. Be careful of the entanglement risks from the gear, cables and line that still cover the deck and much of the wreck.

Other species commonly found at this site:

3 4 10 11 13 14 20 30 31 37 41 44 62 68 72

Name:	*Hydro Atlantic*	**Construction:**	Maryland Steel Corporation, Sparrow Point, Maryland, 1905
Type:	Dredger transport	**Last owner:**	Hydromar Corporation
Previous names:	*Delaware, Sand Captain, Ezra Sensibar*	**Sunk:**	December 7, 1987
Length:	300ft (91.5m)		
Tonnage:	4,276grt		

Wreck Trek Boca

Difficulty ●●○
Current ●●○
Depth ●●○
Reef ★★★
Fauna ★★☆

Access 5 mins from Boca Raton Inlet
20 mins from Hillsboro Inlet

Level Open Water

Location

Boca Raton, Palm Beach County
GPS (*United Caribbean*): 26°19′16.1″N, 80°03′32.3″W

Getting there

As the name suggests, Wreck Trek Boca is located just off shore from Palm Beach County's Boca Raton. The three-wreck trek sits less than a mile off shore, just inside the outer reef. Another site that is shared between the counties, this trek is just a few minutes away from the Boca Raton Inlet, and approximately 20 minutes north of the Hillsboro Inlet. The best way to reach the Boca Wreck Trek is through one of the area's local dive operators.

Access

There is no permanent mooring buoy for the trek, and placement will depend on the local operator as well as the direction of the current. Visibility is generally good, but currents can be strong. Most divers will be able to visit two of the three wrecks on a single dive, typically *United Caribbean* and *Sea Emperor*. The relatively shallow depths of this dive make it suitable for divers with open water certification, although the penetration opportunities make these wrecks appealing for advanced divers as well.

Description

Wreck Trek Boca is a three-wreck trek featuring the *United Caribbean*, *Sea Emperor* and *Noulla Express*. It is a popular trek due to its shallower depth, relative proximity to the inlet, and the quality of the marine life that have colonized the wrecks. The three wrecks are positioned in a triangle. *United Caribbean* and the *Sea Emperor* are located just 400 feet (122 meters) apart with a series of rock piles, known as the Boca Corridors, marking the north-south route between the two wrecks – *Sea Emperor* to the north and *United Caribbean* to the south. *Noulla Express* sits over 500 feet to the east, and as such, most divers typically focus their dive on the other two wrecks. Given that currents tend to run either north or south along the Florida coast, divers often explore both wrecks as part of a drift dive, foregoing the harder to reach *Noulla Express*.

Spadefish school near the bow of the *United Caribbean*.

United Caribbean was originally a 147-foot (45-meter) steel cargo ship built in 1969. Originally named *Golden Venture*, she

RELAX & RECHARGE

If you are looking for great waterfront dining that has an option to suit every palate, then the Waterstone Resort & Marina is probably the place to be if you are in the Boca Raton area. The resort's **Boca Landing** restaurant serves great steak and seafood dishes, including such favorites as seared black grouper and pistachio crusted mahi mahi. You can dine on their large terrace that overlooks the Intracoastal Waterway or in their air-conditioned dining room. For a more casual experience, there is the pool-side **Rum Bar & Grill**. Both establishments are known for their hand-crafted cocktails and the poolside bar often has live music, as well as a happy hour that runs from 4pm to 6pm, Monday to Friday. Visit: **Waterstoneboca.com**

Peter Leahy/Shutterstock ©

operated in Singaporean waters before running aground on Rockaway Beach in Queens, New York in June 1993. She was being used to smuggle Chinese immigrants into the country. She was seized, sold, renamed *United Caribbean*, and began serving as a cargo ship in the Caribbean. She was purposefully sunk as an artificial reef in August 2000. Limited penetration is possible in this wreck, which offers divers a chance to see moray eels, grouper and schools of snapper. Stingrays are often seen in the sand nearby. The wreck sustained some damage during Hurricane Wilma in 2005 and now lies in pieces on a sandy seafloor at a depth of 72 feet (22 meters).

Sea Emperor lies to the north of *United Caribbean*. She was a 171-foot-long (52-meter) hopper barge deliberately sunk as an artificial reef along with

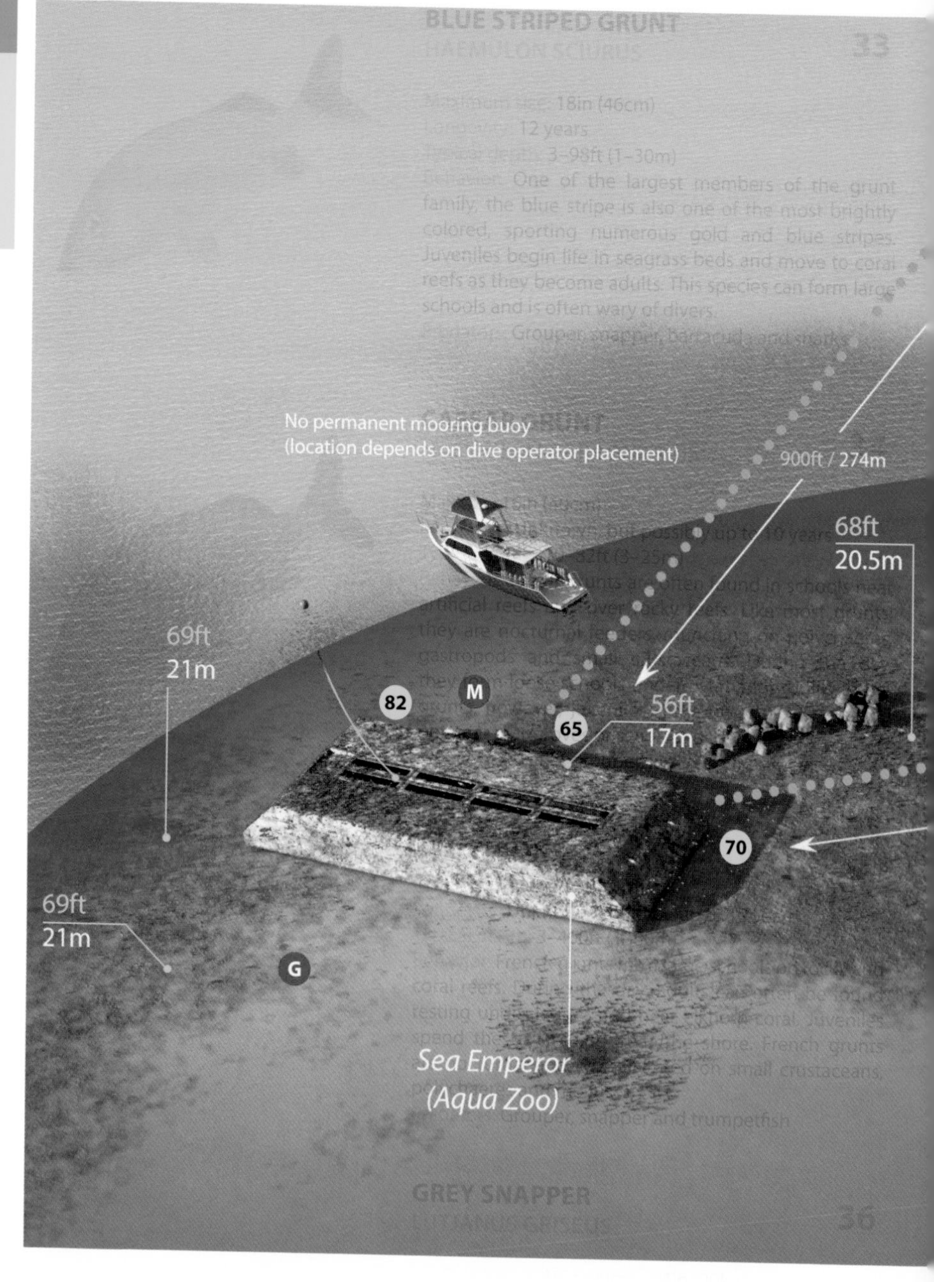

Name: *United Caribbean*
Type: Freighter
Previous names: *Golden Venture*
Length: 147ft (45m)
Tonnage: Unknown
Construction: 1969
Last owner: Unknown
Sunk: August 22, 2000

Name: *Sea Emperor*
Type: Hopper barge
Previous names: n/a
Length: 171ft (52m)
Tonnage: Unknown
Construction: Unknown
Last owner: A dredging company that damaged the reef
Sunk: October 1991

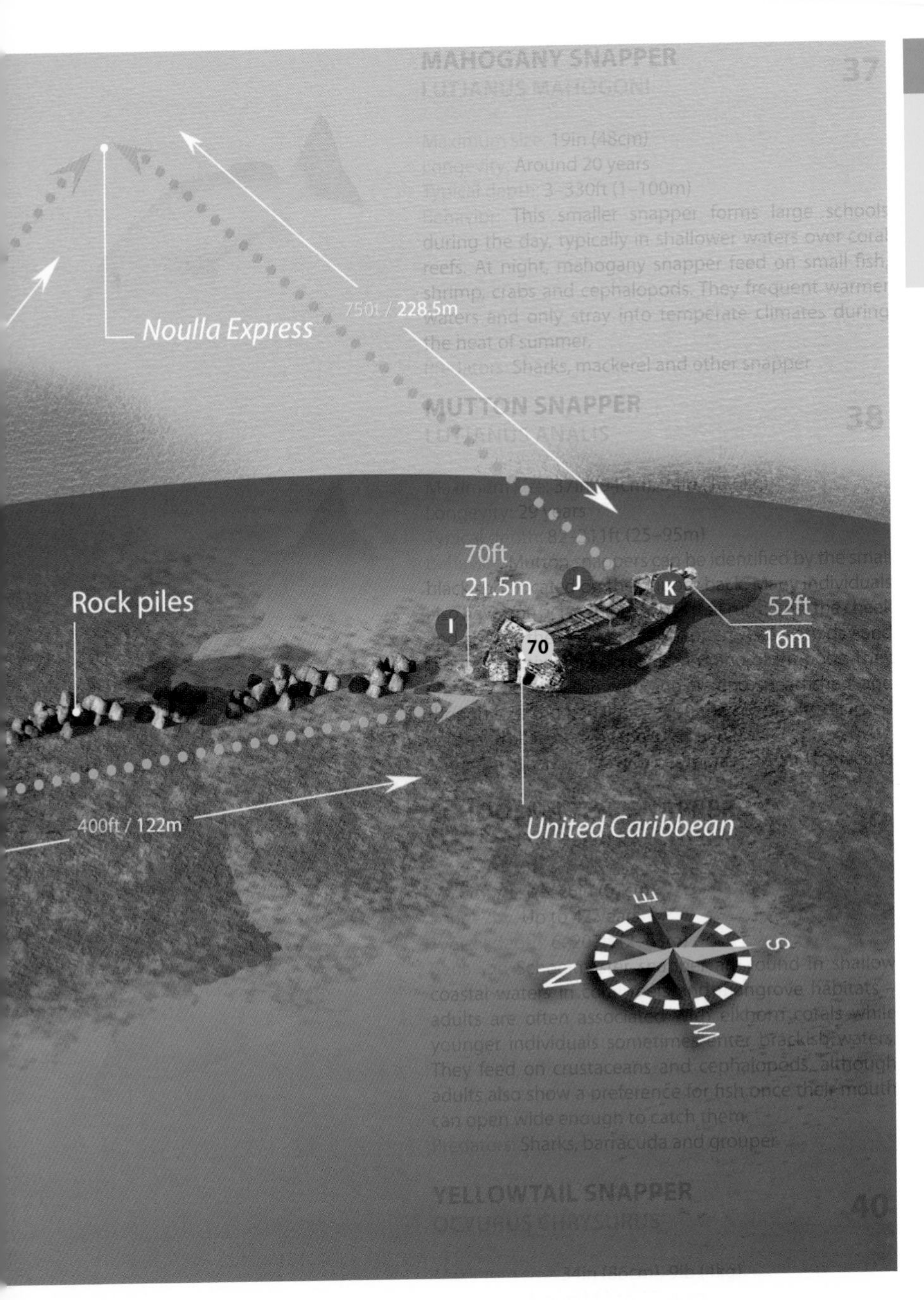

Name:	*Noulla Express*
Type:	Freighter
Previous names:	*Danaland, Trean, Kornmod, Kormoran, Cantadora, La Verdad*
Length:	114ft (34.5m)
Tonnage:	174grt
Construction:	Denmark 1931
Last owner:	Unknown
Sunk:	July 12, 1988

Other species commonly found at this site:

1 2 6 7 9 10

24 25 41 43

1,600 tons of concrete drainage culverts. She was donated by a dredging company that had damaged a section of reef in Palm Beach County. *Sea Emperor* tipped over as she sank, settling upside down and spilling the culverts onto the nearby sandy bottom. The resulting artificial reef is known as the *Aqua Zoo* by some dive operators because of the great variety of organisms that can be found in this marine playground. A resident green moray and nurse shark are often found at this site, as well as the occasional hammerhead shark. The wreck can be safely penetrated via multiple access points, and light filters into all chambers. The overturned hull and culverts are home to a multitude of marine organisms, and

UNITED CARIBBEAN

United Caribbean position

are covered in encrusting corals and sponges.

Noulla Express lies to the east of the other two wrecks in this trek, just beyond a number of patch reefs. She was the first artificial

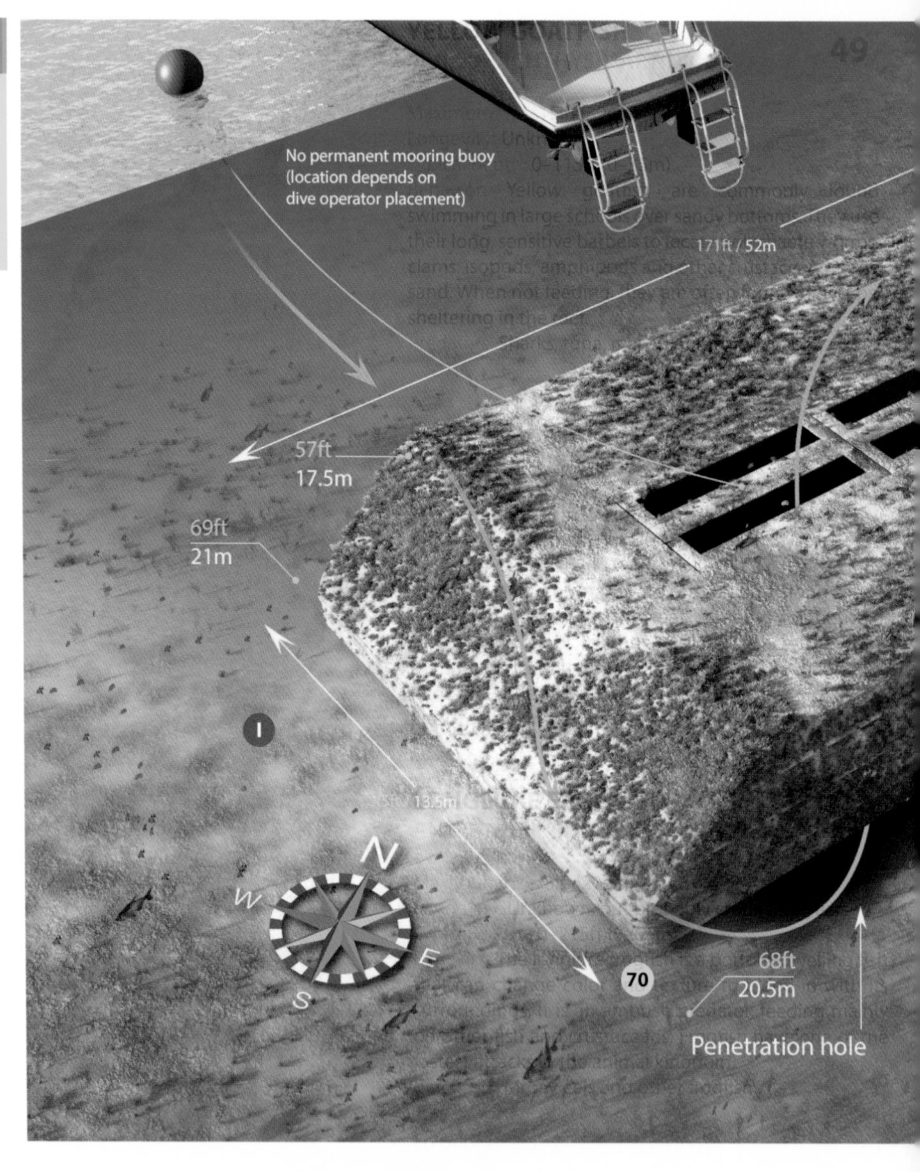

reef jointly sunk by the Broward and Palm Beach counties. Originally a Danish freighter built in 1931, *Noulla Express* changed names and owners multiple times over her decades of service in Scandinavia before moving to this side of the Atlantic under a Panamanian registration. She was seized by the Federal government in a drug case and jointly purchased by the artificial reef programs of the two counties. Workers used blow torches to cut holes in her aging hull and she was scuttled in July of 1988. Hurricane Andrew did some damage in 1992, tearing her into three pieces, tilting her bow upwards and flattening her cargo holds. While the years have not been kind to the wreck, they have provided corals and sponges time to completely colonize her surfaces. She now plays host to a huge diversity of marine life, particularly filefish and angelfish.

Route

The best route will depend on the currents and whether divers want to attempt exploring all three sites. The *United Caribbean* and *Sea Emperor* are certainly the easiest to explore and navigate thanks to the location of the Boca Corridors rock piles. As a drift dive,

Sea Emperor position

divers will descend on the first wreck, whichever is upstream with respect to the current, and circumnavigate this wreck before following the rock piles to the second wreck. Penetration is possible on both wrecks for those divers with the experience and necessary certification. In particular, *Sea Emperor* offers plenty of prepared access points for divers with the necessary qualifications.

Ancient Mariner

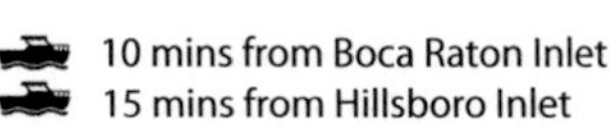

Level Open Water

Location

Deerfield Beach, Broward County
GPS (*Ancient Mariner*): 26°18'07.3"N, 80°03'43.7"W

Getting there

Ancient Mariner is part of a three-wreck trek located just over a mile off shore from Deerfield Beach in Broward County. It is a 10-minute ride from the Boca Raton Inlet and about 15 minutes north of the Hillsboro Inlet. The trek consists of three wrecks arranged in a triangle, with the two main wrecks positioned relatively close to one another and a third located 600 feet (183 meters) to the south. The trek sits just inside the outer reef, and the best way to reach it is through one of the local dive operators active out of either the Boca Raton or Hillsboro Inlets.

Access

There is no permanent mooring buoy for the *Ancient Mariner* or the Deerfield Wreck Trek, and placement will depend on the local operator as well as the direction of the current. Visibility is generally good, but currents can be strong. In the case of a steady current, operators may choose to drop divers into the water without mooring, in order to let them experience the trek as a drift dive – be sure to have your surface

marker buoy with you. Operators may choose to tie up to the *Ancient Mariner*, the main attraction of the trek, and limit the dive to just the two wrecks positioned at the northern end of the trek. The shallower depths of this dive make it suitable for divers with Open Water certification.

Description

Ancient Mariner is part of the Deerfield Wreck Trek, which also features the *Berry Patch Tug* and *Quallman Barge* (also *Qualman Barge*). This trek is popular due to its relatively shallow depth and the quality marine life that have colonized the artificial reefs. It is also popular for divers looking to practice their navigational skills, as *Quallman Barge* rests 600 feet (183 meters) to the south of the other two wrecks. All three wrecks were purposefully sunk as artificial reefs at a depth of 65 to 75 feet (20 to 23 meters).

Ancient Mariner was originally a B-class Coast Guard cutter named *USCG Nemesis*. She launched in 1934 to assist in the fight against Prohibition-era rum-running from the Bahamas to the United States, although she launched just after Prohibition ended. When World War II started, *Nemesis* was transferred to the Navy to serve as a Nazi U-boat chaser and convoy escort based out of St. Petersburg, Florida. The Navy transferred her back to the Coast Guard at the end of the war, and she was eventually decommissioned in 1964.

Investors purchased the ship in 1979 and converted her into a floating restaurant, remodeling her upper decks to give her the look of an African steamer. They renamed her *Livingstone's Landing*, after the nineteenth century explorer and medical missionary Dr. David Livingstone. The venture failed, and the ship was relaunched several more times as a floating restaurant, under various names and owners, but none of the businesses succeeded.

In June 1991, *Ancient Mariner* was sunk as an artificial reef off Deerfield Beach as part of a joint venture by the Broward County and Palm Beach County artificial reef programs. She settled on the seabed at a depth of 70 feet (21.5 meters) with her bow facing southeast. Below deck, the wreck has plenty of opportunities for penetration for those with the necessary qualifications and experience. Above deck, a small two-level structure initially provided a swim-through opportunity for divers. However, the first above-deck level of the *Ancient Mariner* started to collapse toward the port side during

DID YOU KNOW?

The owners of the Livingstone's Landing floating restaurant placed the air-conditioned kitchens on the middle deck, exhibiting them for the benefit of the vessel's diners long before this became a common practice in the restaurant industry.

The restaurant closed in 1981 and partially sank while in port and tied to the dock. The ship was re-floated and repaired at a cost of more than $85,000 and relaunched as another floating restaurant, this time named the *Ancient Mariner*. In 1986, a kitchen worker spread hepatitis A to more than 100 people – the largest food-borne outbreak of hepatitis in Florida history to date. The restaurant closed in the wake of the outbreak and the owners filed for bankruptcy.

The wheelhouse of the *Berry Patch Tug*, which is part of the same wreck trek as the *Ancient Mariner*.

Peter Leahy/Shutterstock ©

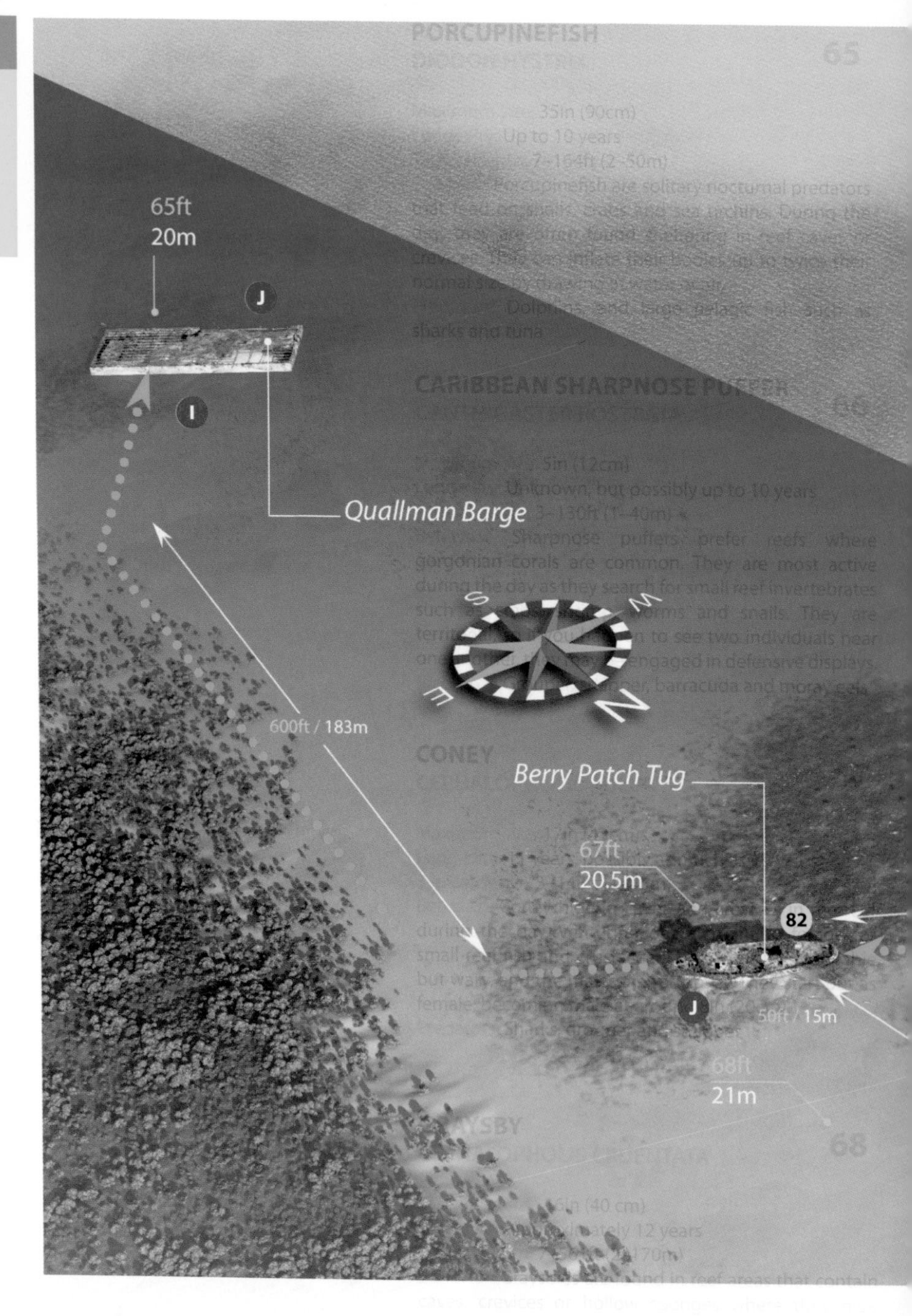

Hurricane Sandy in 2012, causing the second level to partially collapse as well. Hurricane Irma finished the job started by Sandy, ripping what remained of the superstructure off the wreck and dumping it on the nearby sand in 2017.

Route

The most common route will have divers descend onto *Ancient Mariner*. Depending on the strength of the current, divers can easily circumnavigate the wreck, pausing to check out the collapsed superstructure on the port

* Other species commonly found at this site:  K 2 26 27 36 40 63 73 76 88

side. Follow the rebar stakes off the bow of the wreck to find *Berry Patch Tug* that lies just 150 feet (45.5 meters) away on the sandy seafloor. Divers should not forget to glance to the left to catch a glimpse of the *Mariner*. They can circumnavigate *Berry Patch* looking inside the wheelhouse to view the encrusting corals and sponges that have colonized the wreck. If the current allows, continue the dive toward the outer reef, before turning right on a southern heading that parallels the reef for roughly 450 feet. Next, turn to a 240-degree heading to

Ancient Mariner position

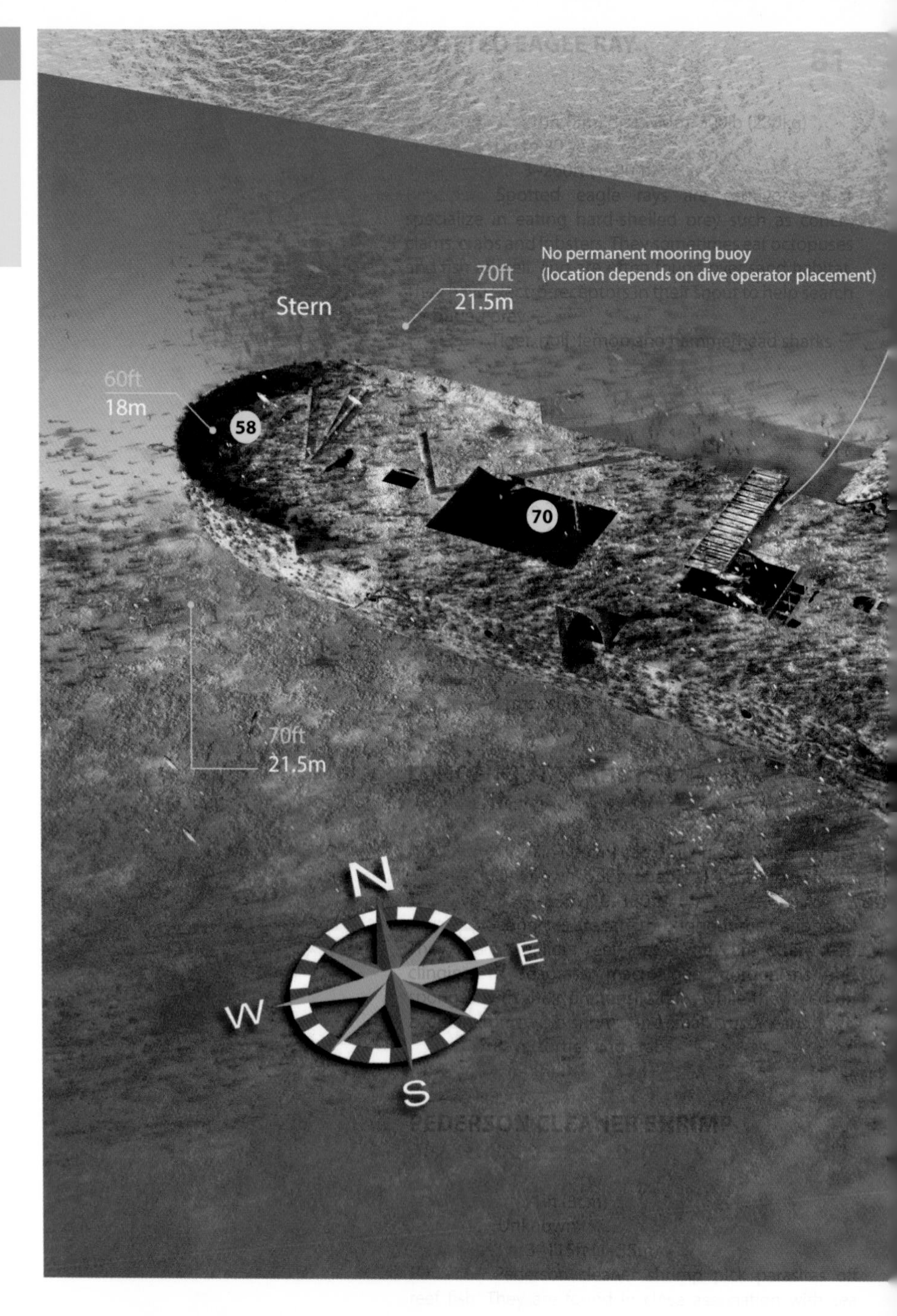

follow a set of rebar stakes buried in the sand to reach the *Quallman Barge*.

Name:	*Ancient Mariner*	**Last owner:**	United States government
Type:	USCG cutter	**Sunk:**	June 9, 1991
Previous names:	*Nemesis*		
Length:	165ft (50.5m)		
Tonnage:	337grt		
Construction:	Point Pleasant, West Virginia, 1934		

Species

Identifying coral reef organisms is an enjoyable part of any underwater adventure. Not only can you appreciate the diversity and wonder that surrounds you on a reef, but you will be better able to understand the story that is unfolding right before your eyes.

For example, you will know where and when to look for certain species, as well as what they eat, who eats them, how big they get and how long they live. But more specifically, you will understand certain behaviors that can be observed on coral reefs, such as why damselfish attack larger creatures or which creatures form symbiotic relationships and why.

Many times, behaviors are an integral part of the identification process. In some cases, understanding how a particular fish behaves,

SAFETY TIP

The section on dangerous species that follows is intended to provide the information you need to recognize the handful of species that can cause injury. These species should not be considered "active threats," but rather organisms that have the potential to cause harm. Most injuries occur because the organism in question has felt threatened and because a diver or snorkeler has not recognized the warning signs. By engaging in safe and conscientious diving and snorkeling practices, and by keeping in mind a few key safety tips, you can avoid having your experience ruined by an unpleasant sting or bite.

Hogfish are common throughout Palm Beach County.

such as whether it is active during the night or day or whether it is an ambush predator or active forager, can be more useful in determining its identity than its color or shape.

Many reef organisms may appear very similar at first glance, and the wide diversity of species on coral reefs can appear to be a chaotic jumble. But by combining an understanding of animal behavior with some basic identification information, you can start to tease apart that puzzle and experience the wonder of the coral reef.

Ethan Daniels/Shutterstock ©

A Caribbean spiny lobster clambers from under a reef ledge.

megablaster/Shutterstock ©

The information provided in this guide represents the most up-to-date science available at the time of publication. It covers some of the most common reef species you will find during your time in the waters off Palm Beach County, Florida However, it should be noted that coral reef ecologists continue to discover new information about species, their behaviors and their interactions. Later editions of this book may contain modifications that reflect new knowledge.

The following pages are divided into three sections that feature information about sea turtles, which have a rather unique life history; dangerous species, including details on the kind of threat they pose and how to treat injuries caused by them; and finally, a general species section that helps you identify and learn about the most common species found at the dive and snorkel sites featured in this guide.

Sea turtle identification

A

GREEN TURTLE
CHELONIA MYDAS

Maximum size: 4ft (1.4m)
Longevity: Up to 75 years
Habitat: Seagrass beds, reefs
Diet: Jellyfish and crustaceans when young; algae and seagrass as adults
Sightings: Common
Behavior: Green turtles can be found grazing on vegetation in shallow water or cruising the reef. Most green turtles migrate short distances along the coast to reach nesting beaches, but some may migrate up to 1,300 miles (2,100 kilometers) to reach nesting beaches.
Predators (adults): Tiger sharks, orcas (killer whales)

B

HAWKSBILL SEA TURTLE
ERETMOCHELYS IMBRICATA

Maximum size: 3ft (0.9m)
Longevity: Up to 50 years
Habitat: Reefs
Diet: Sponges, tunicates, squid, shrimp
Sightings: Common
Behavior: Hawksbills can be found feeding throughout the day or resting with their bodies wedged into reef cracks and crevices. Some hawksbills do not migrate at all, while others migrate over thousands of miles.
Predators (adults): Tiger sharks, orcas (killer whales)

SEA TURTLE CONSERVATION STATUS: ENDANGERED

All species of sea turtles are endangered (Kemp's ridleys are critically endangered) and many human activities contribute to their decline.

- Turtles are hunted in many parts of the world, targeting their meat and eggs for food, and their shells to make jewelry, eyeglass frames and curios. Many others drown in fishing nets intended for shrimp or fish, or are struck and killed by passing boats.
- Turtles eat and choke on plastic and other trash, while pollution increases the frequency of turtle disease.
- Coastal development is rapidly reducing the number of active nesting beaches.

ECO TIP

Sea turtles need our help to survive and thrive alongside our coastal communities. Consider the following tips:

LIGHTS OUT
Turn out lights visible from the beach to avoid disorienting nesting turtles and hatchlings.

DON'T LITTER
Plastic cups, bags and other trash can kill turtles when they mistake them for food.

DON'T DISTURB
Turtles you see on the beach or in the water, whether nesting, feeding or hatching, should be left alone and observed from a distance – without flashlights.

PLEASE DON'T FEED
Human food can make sea turtles sick and can leave them vulnerable to capture.

VOLUNTEER
Support your local turtle conservation programs, including participating in beach cleanups.

LOGGERHEAD SEA TURTLE
CARETTA CARETTA

C

Maximum size: 3.5ft (1.1m)
Longevity: Up to 60 years
Habitat: Reefs and open ocean
Diet: Crabs, shrimp, jellyfish, vegetation
Sightings: Common
Behavior: Loggerheads are occasionally found in the open ocean, but regularly move inshore to feed on reef invertebrates. Loggerhead sea turtles can migrate for thousands of miles to reach new feeding grounds before returning to the same nesting beaches.
Predators (adults): Tiger sharks, orcas (killer whales)

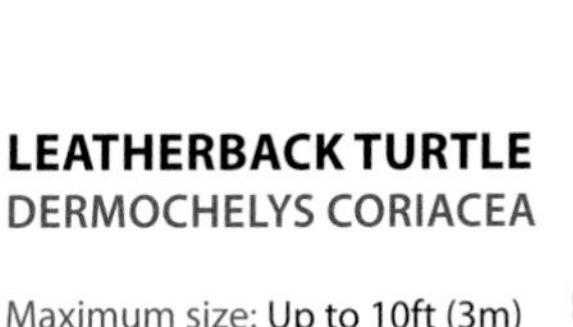

LEATHERBACK TURTLE
DERMOCHELYS CORIACEA

D

Maximum size: Up to 10ft (3m)
Longevity: Up to 80 years
Habitat: Open ocean
Diet: Jellyfish and tunicates
Sightings: Rare
Behavior: Leatherbacks feed in deep water during the day and at the surface at night, following the daily migratory patterns of their favorite food: jellyfish. During the mating season, leatherbacks may migrate up to 3,000 miles (4,800 kilometers) from their feeding grounds to their nesting beaches.
Predators (adults): Tiger sharks, orcas (killer whales)

KEMP'S RIDLEY SEA TURTLE
LEPIDOCHELYS KEMPII

E

Maximum size: Up to 2ft (0.7m)
Longevity: Up to 50 years
Habitat: Nearshore, shallower water
Diet: Crabs, shellfish, jellyfish and small fish
Sightings: Rare
Behavior: Kemp's ridleys are the smallest of the Florida sea turtles, and the only one that nests primarily during the day – 95 percent of their nesting activity takes place in Mexico. They practice mass nesting, where thousands of females come ashore at the same time to lay eggs. Individuals inhabit shallow, coastal waters, using their large, triangular crushing beak to feed on their favorite food: crabs.
Predators (adults): Tiger sharks, orcas (killer whales)

Sea turtle ecology

Nesting
Females crawl onto the beach at night (or during the day in the case of Kemp's ridleys) and dig a shallow nest. They lay up to 200 small white eggs before covering the nest and returning to the sea. The eggs incubate for 45 to 70 days, depending on the species.

Mating
Most sea turtles reproduce in the warm summer months except for the leatherback, whose mating season spans fall and winter. Many species migrate great distances to return to the their customary nesting beach. Courtship and mating occur in the shallow waters off shore.

1

Pete Niesen/Shutterstock ©

5

Gail Johnson/Shutterstock ©

Adulthood
In adulthood, some sea turtle species return to coastal waters where coral reefs and nearshore waters provide plenty of food and protection from predators.

UWPhotog /Shutterstock ©

Matt Jeppson/Shutterstock ©

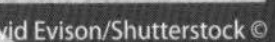

David Evison/Shutterstock ©

BlueOrange Studio/Shutterstock ©

Hatching

At hatching, hundreds of tiny turtles dig their way out of the nest and head toward the moonlight reflecting off the sea. As many as 90 percent are eaten by predators as eggs or as hatchlings within the first few hours of their lives.

Willyam Bradberry/Shutterstock ©

Juvenile stage

Young turtles drift through the open ocean for years, often associating with floating sargassum (seaweed) mats. They feed on plankton and small jellyfish. Little is known about this stage of their lives.

Dangerous species

MEDICAL DISCLAIMER

The treatment advice contained in this book is meant for informational purposes only and is not intended to be a substitute for professional medical advice, either in terms of diagnosis or treatment. Always seek the advice of your physician or other qualified health provider if you are injured by a marine organism. Never disregard professional medical advice or delay seeking it because of something you have read in this book.

SCALLOPED HAMMERHEAD SHARK

SPHYRNA LEWINI

Maximum size: 14ft (4.3m), 990lb (449kg)
Longevity: About 35 years
Typical depth: 3–984ft (1–300m)
Behavior: Scalloped hammerhead sharks are found in tropical and warm temperate marine waters. Their name comes from the scalloped grooves in their hammer-shaped snout. They specialize in hunting stingrays, but also feed on grouper, snapper, and other shark species.
Predators: Larger shark species and orcas (killer whales)

LEMON SHARK

NEGAPRION BREVIROSTRIS

Maximum size: 11ft (3.4m), 405lb (183kg)
Longevity: About 30 years
Typical depth: 3–302ft (1–92m)
Behavior: Lemon sharks are found in the Atlantic and Pacific Oceans. They are lighter than most sharks, which helps them blend into sandy backgrounds. They most often feed at night on a range of fish and crustaceans. Larger lemon sharks may even eat smaller lemon sharks.
Predators: Larger shark species, such as bull and tiger sharks

CARIBBEAN REEF SHARK

CARCHARHINUS PEREZII

Maximum size: 10ft (3m) (10ft)
Longevity: Unknown, but possibly 20 years or more
Typical depth: 3–213ft (1–65m)
Behavior: The Caribbean reef shark is the most common shark species encountered on Caribbean coral reefs. They tend to inhabit drop-offs and the seaward edges of reefs where they feed on most reef fish species, as well as stingrays and eagle rays.
Predators: Larger shark species, such as bull and tiger sharks

WARNING FOR THESE SHARKS: Attacks on humans by lemon sharks, hammerheads and Caribbean reef sharks are rare, but they have been known to cause injury if threatened or cornered. Warning signs of an attack include head swings, exaggerated swimming, back arching and lowered pectoral fins. Attacks usually result in biting or raking with the teeth, which can cause deep lacerations.

TREATMENT: Exit the water as soon as possible and rinse the affected area with soap and water. Apply pressure to control the bleeding and elevate the affected limb above the heart. Shark bite victims sometimes require treatment for shock, in which case, keep them warm, calm and in the shade, and do not provide anything to eat or drink. Lay them on their back and elevate the legs above the head. Seek medical attention as soon as possible, even for minor bites, for proper cleaning and suturing.

SOUTHERN STINGRAY
DASYATIS AMERICANA

I

Maximum size: 7ft (2m) disk diameter, 300lbs (136kg)
Longevity: Unknown, but probably over 10 years
Typical depth: 0–170ft (0–53m)
Behavior: Stingrays are most active at night when they hunt for hard-shelled prey such as snails, crabs, lobsters and occasionally fish. During the day, they are often found buried up to their eyes in sand.
Predators: Sharks and large grouper

WARNING: Stingrays have a serrated venomous spine at the base of their tail that they use for defense. The area around a puncture wound from this spine may become red and swollen, and you may experience muscle cramps, nausea, fever and chills.

TREATMENT: If you are stung, exit the water immediately. Apply pressure above the wound to reduce bleeding, clean the wound and soak the area with hot water, ideally around 113°F (45°C), to reduce the pain. Apply a dressing and seek medical attention. Antibiotics may be needed to reduce the risk of infection. Stingray injuries can be very painful, often reaching a peak around one hour after the injury and lasting up to two days. But they are rarely fatal unless the injury is to the head, neck or abdomen.

GREAT BARRACUDA
SPHYRAENA BARRACUDA

Maximum size: 6ft (2m), 110lbs (50kg)
Longevity: Around 20 years
Typical depth: 1–100m (3–330ft)
Behavior: Barracuda are most active during the day, feeding on jacks, grunts, grouper, snapper, squid and even other barracuda. They are often solitary in nature, but occasionally school in large numbers. They have even been documented "herding" fish they plan on consuming. Barracuda use their keen eyesight to hunt for food. They are one of the fastest fish in the ocean, capable of bursts of speed up to 30mph (48kph). Along with their two sets of razor sharp teeth, there are few prey capable of escaping a barracuda once it decides to attack.
Predators: Sharks, tuna and large grouper

WARNING: Barracuda do not usually attack divers or snorkelers unless provoked. However, evidence suggests they are attracted to objects that glint or shine, such as necklaces, watches or regulators, which they may mistake for prey. The bite of the barracuda is not toxic, but their teeth can produce a severe laceration or deep puncture wound.

TREATMENT: Exit the water as soon as possible and apply pressure to reduce bleeding. The wound should be cleaned and dressed. Medical attention may be necessary for severe bites, including sutures to close the wound and antibiotics to reduce the risk of infection.

SPOTTED SCORPIONFISH
SCORPAENA PLUMIERI

Maximum size: 18in (45cm)
Longevity: Around 15 years
Typical depth: 3–197ft (1–60m)
Behavior: Spotted scorpionfish spend much of their time lying motionless on the seabed, using camouflage to ambush fish and crustaceans. They have a large, expandable mouth capable of creating a vacuum to suck in prey, which they swallow whole.
Predators: Large snapper, sharks, rays and moray eels

WARNING: Scorpionfish have a dozen venomous dorsal spines for self-defense. The spines can penetrate skin (most commonly when stepped on), injecting a toxin that causes severe pain that can last from several hours to several days. The area around the injury may also swell and become red.

TREATMENT: Exit the water quickly and rinse the affected area with seawater. Remove any spines and use pressure to control any bleeding. Apply the hottest water you can stand to reduce pain, ideally around 113°F (45°C). Let the wound heal uncovered, but antibiotics may be required to avoid infection. The toxin can be painful, but it is not usually fatal. However, seek medical attention if concerned or if symptoms are worse than described.

RED LIONFISH

PTEROIS VOLITANS

Maximum size: 15in (38cm)
Longevity: Around 10 years
Typical depth: 7–180ft (2–55m)
Behavior: Red lionfish are originally from the Indo-West Pacific, and are considered an invasive species in the Western Atlantic. They are most active at dusk and during the night when they hunt for fish, shrimp, crabs and other reef creatures. Lionfish can live without food for up to three months.
Predators: Occasional predation by certain sharks and grouper

WARNING: Lionfish have up to 16 venomous dorsal and anal spines that can deliver a powerful neurotoxin when they puncture skin. Lionfish do not generally attack divers and snorkelers, but may sting in self-defense if you get too close. Divers and snorkelers may feel intense pain after being stung, followed by swelling and redness around the wound.

TREATMENT: Exit the water as soon as possible and remove any pieces of the spines that may remain in the wound. Use pressure to control the bleeding and apply the hottest water you can stand, ideally around 113°F (45°C), to reduce the pain. Some people experience shortness of breath, dizziness and nausea. There have been no known fatalities from a lionfish sting, but there is always a risk of complications for vulnerable individuals, including congestive heart failure. Seeking medical attention is advised. The pain may last anywhere from several hours to several days.

GREEN MORAY EEL

GYMNOTHORAX FUNEBRIS

Maximum size: 8ft (2.5m), 65lbs (30kg)
Longevity: Unknown
Typical depth: 3–164ft (1–50m)
Behavior: Green morays are solitary animals that hide in reef cracks and crevices during the day. At night, they prey on fish, octopuses, crustaceans and even other eels, primarily using smell to hunt as their eyesight is poor.
Predators: Unknown

WARNING: Moray eels have sharp teeth that can produce a painful wound, but thankfully they rarely attack unless provoked. There is evidence that the bite of some morays may contain toxins that increase pain and bleeding, but more research is needed. Although all morays can bite, larger species such as the green moray eel can cause more severe injuries than smaller species.

TREATMENT: Exit the water as soon as possible. Treat the wound by immediately cleaning the affected area with soap and water. Apply pressure to reduce the bleeding, then apply a topical antibiotic before dressing the wound to reduce the risk of infection. Sutures may be required in some cases. If in doubt or if the wound becomes infected, seek medical attention.

JELLYFISH & SIPHONOPHORES

HYDROIDOMEDUSAE & SIPHONOPHORAE

Maximum size: 7ft (2m) with tentacles extending much farther
Longevity: From a few hours to several years
Typical depth: 0–66ft (0–20m)
Behavior: Jellyfish and siphonophores are both types of cnidaria. Jellyfish are individual animals, while siphonophores are colonies of specialized cells called zooids, such as the Portuguese man o' war. They both drift in the water and use stinging cells called nematocysts to capture and paralyze prey, including plankton and small fish.
Predators: Salmon, tuna and some sharks and sea turtle species

WARNING: Jellyfish and siphonophores have stinging cells called nematocysts located on their tentacles that can inject a toxin when brushed against bare skin. Depending on the species, jellyfish toxin can cause mild tingling to intense pain, and can be fatal in some rare cases. The contact site may also become red and blistered. Even dead jellyfish on the beach can sting, so avoid touching them.

TREATMENT: Exit the water as quickly as possible, watching out for other jellyfish. Rinse the affected area with seawater to remove any pieces of tentacle on the skin. Do not rinse with fresh water, which can trigger any remaining nematocysts to sting. The best treatment for jellyfish stings may depend on the species, but most can be treated by rinsing the affected area with vinegar or creating a paste using baking soda and seawater. The papain enzyme found in meat tenderizer and papaya can also help. Consider seeking medical attention.

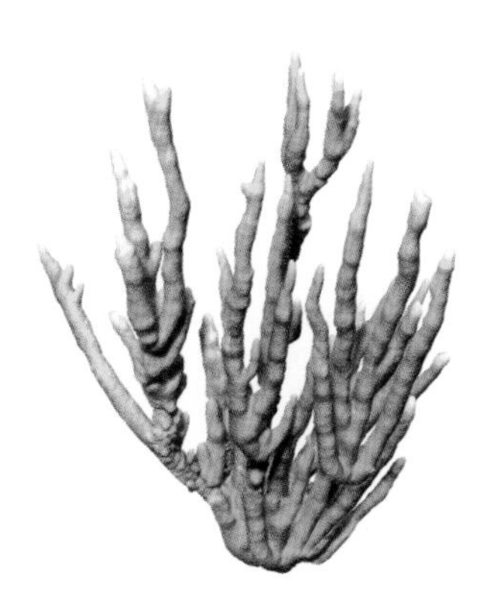

FIRE CORAL

MILLEPORIDAE

Maximum size: 3ft (1m)
Longevity: Unknown, but likely decades
Typical depth: 3–130ft (1–40m)
Behavior: Several fire coral species occur in the Caribbean, attaching to the reef substrate and growing in branching, blade and encrusting forms. Fire corals are hydroids with a hard skeleton, and are more closely related to jellyfish than corals. They get their energy from photosynthetic zooxanthellae in their tissues, but also from feeding on plankton.
Predators: Fireworms, certain nudibranchs and filefish

WARNING: Microscopic fire coral polyps are located throughout the surface of the hard skeleton. Each polyp has hair-like tentacles that are covered in stinging cells called nematocysts, which they use to paralyze their tiny prey. Fire corals cause a lingering, burning sensation when they contact bare skin. A rash or blistering may occur in some individuals and may last several days, but is not usually dangerous.

TREATMENT: Exit the water as soon as possible and rinse the affected area with vinegar or alcohol to inactivate the fire coral toxin. Do not rinse with fresh water, which can increase the pain by causing untriggered nematocysts to discharge into the skin. Apply hydrocortisone cream to the area once dry. The papain enzyme found in meat tenderizer and papaya can also reduce swelling, pain and itching.

FRENCH ANGELFISH
POMACANTHUS PARU

1

Maximum size: 24in (60cm)
Longevity: Up to 15 years
Typical depth: 10–330ft (3–100m)
Behavior: French angelfish dine primarily on sponges, but may also feed on gorgonians and algae. Juveniles often act as cleaners, eating the parasites from other reef fish. At dusk, French angelfish find shelter from nocturnal predators in reef cracks and crevices.
Predators: Large grouper and sharks

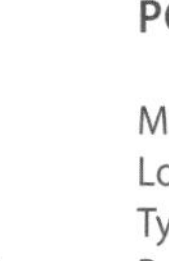

GRAY ANGELFISH
POMACANTHUS ARCUATUS

2

Maximum size: 24in (60cm)
Longevity: Unknown, possibly up to 15 years
Typical depth: 6–100ft (2–30m)
Behavior: Gray angelfish are often seen swimming in pairs as they are known to form long-term, monogamous breeding pairs. They are recognizable by their grey-brown bodies and pale grey-white mouths. They frequent coral reefs, feeding on sponges, tunicates, hydroids, algae and sometimes seagrass.
Predators: Large grouper and sharks

BLUE ANGELFISH
HOLOCANTHUS BERMUDENSIS

3

Maximum size: 18in (45cm)
Longevity: Unknown, possibly up to 15 years
Typical depth: 6–300ft (2–91m)
Behavior: Blue angelfish are often mistaken for queen angelfish, but they lack the distinct forehead crown and are paler in color. Like most angelfish, they are often seen swimming in pairs, foraging on sponges and small benthic invertebrates. At night, they sleep hidden away in the reef, safe from predators.
Predators: Large grouper and sharks

QUEEN ANGELFISH
HOLACANTHUS CILIARIS

Maximum size: 18in (45cm)
Longevity: Up to 15 years
Typical depth: 3–230ft (1–70m)
Behavior: Queen angelfish are often found swimming gracefully between seafans, sea whips and corals, alone or in pairs. They feed almost exclusively on sponges, but have been known to snack on algae and tunicates as well. Young Queen angelfish also clean parasites off larger fish.
Predators: Large grouper and sharks

ROCK BEAUTY
HOLACANTHUS TRICOLOR

5

Maximum size: 14in (35cm)
Longevity: Up to 20 years (in captivity)
Typical depth: 10–115ft (3–35m)
Behavior: Adult rock beauties are often found on rock jetties, rocky reefs and rich coral areas, while juveniles tend to be found near fire corals. These angelfish are not picky eaters and will feed on tunicates, sponges, zoantharians and algae.
Predators: Grouper, snapper and sharks

BLUE TANG
ACANTHURUS COERULEUS

6

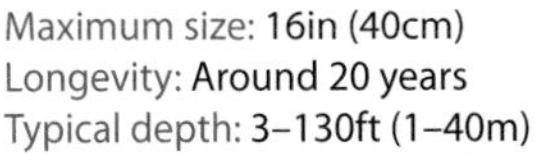

Maximum size: 16in (40cm)
Longevity: Around 20 years
Typical depth: 3–130ft (1–40m)
Behavior: Blue tangs are often found grazing on algae during the day, either individually or as part of large schools that may also contain surgeonfish, doctorfish, goatfish and parrotfish. At dusk, they settle into a reef crack or crevice to hide for the night.
Predators: Grouper, snapper, jacks and barracuda

DOCTORFISH
ACANTHURUS CHIRUGUS

7

Maximum size: 15.5in (39cm)
Longevity: Up to 30 years
Typical depth: 6–213ft (2–65m)
Behavior: Doctorfish can be found in shallow, inshore reef habitats and rocky areas. They forage on benthic algae, including the thin algal mat covering sandy bottoms. They generally swim together in loose schools, often with ocean surgeonfish and blue tangs. They have sharp spines near their tail fin that they can use in defense against predators.
Predators: Large carnivorous fish, including tuna

OCEAN SURGEONFISH
ACANTHURUS BAHIANUS

8

Maximum size: 15in (38cm)
Longevity: Up to 32 years
Typical depth: 6–130ft (2–40m)
Behavior: Adult surgeonfish often form large schools to graze on benthic algae and seagrasses in shallow coral reefs and inshore rocky areas. Juveniles rarely school, sheltering instead in the back reef. Researchers have observed spawning aggregations of up to 20,000 individuals in the winter months off of Puerto Rico.
Predators: Sharks, grouper, barracuda and snapper

BANDED BUTTERFLYFISH
CHAETODON STRIATUS

9

Maximum size: 6in (16cm)
Longevity: Unknown, but probably around 10 years
Typical depth: 10–60ft (3–20m)
Behavior: Banded butterflyfish are most active during the day when they search the reef for food, which includes polychaete worms, zoanthids, anemones and fish eggs. Banded butterflyfish are often found in monogamous pairs and they defend a joint territory together with their mate.
Predators: Moray eels and large carnivorous fish

FOUREYE BUTTERFLYFISH
CHAETODON CAPISTRATUS

10

Maximum size: 6in (15cm)
Longevity: Around 8 years
Typical depth: 6–65ft (2–20m)
Behavior: Foureye butterflyfish are active during the day when they feed on small invertebrates. Their pointed mouth allows them to pull prey from small crevices. They are often found in pairs, and males and females bond early in life and form long-lasting monogamous pairs.
Predators: Barracuda, grouper, snapper and moray eels

REEF BUTTERFLYFISH
CHAETODON SEDENTARIUS

11

Maximum size: 6in (15cm)
Longevity: Unknown, but probably around 10 years
Typical depth: 16–302ft (5–92m)
Behavior: This species is one of the deepest dwelling Caribbean butterflyfish. Like many members of this family, their color and pattern disguise the head in an attempt to confuse potential predators. Reef butterflyfish are most active during the day when they feed on polychaete worms and small crustaceans. They particularly like to eat the eggs of sergeant majors.
Predators: Barracuda, grouper, snapper and moray eels

SPOTFIN BUTTERFLYFISH
CHAETODON OCELLATUS

12

Maximum size: 8in (20cm)
Longevity: Unknown, but probably around 10 years
Typical depth: 3–98ft (1–30m)
Behavior: The spotfin butterflyfish can be identified by the small spot on the rear end of the dorsal fin. This species is found over an incredibly large geographical area, extending from southern Brazil to as far north as Nova Scotia, Canada. Spotfin butterflyfish are most active during the day when they search for food, which includes polychaete worms, zoanthids, anemones and fish eggs.
Predators: Barracuda, grouper, snapper and moray eels

BLUE CHROMIS
CHROMIS CYANEA

13

Maximum size: 5in (12cm)
Longevity: Unknown, possibly 5 years
Typical depth: 10–70ft (3–20m)
Behavior: Blue chromis gather in schools above the reef to feed on small plankton and jellyfish during the day. They hide in reef crevices at night. Territorial males defend egg nests in the spring and summer.
Predators: Trumpetfish, grouper and snapper

BROWN CHROMIS
CHROMIS MULTILINEATA

14

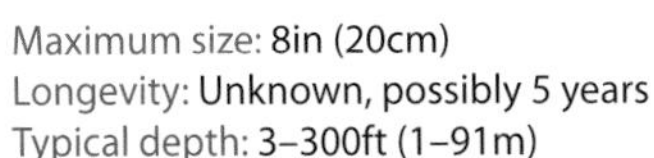

Maximum size: 8in (20cm)
Longevity: Unknown, possibly 5 years
Typical depth: 3–300ft (1–91m)
Behavior: Brown chromis forage in medium-sized schools above the coral reef, feasting on plankton, mainly copepods. They are frequently seen schooling with blue chromis during the day, although their more territorial congeneric tends to chase them out from hiding places in the reef at night.
Predators: Trumpetfish, grouper and snapper

SERGEANT MAJOR
ABUDEFDUF SAXATILIS

15

Maximum size: 9in (23cm)
Longevity: Unknown, possibly 5 years
Typical depth: 3–33ft (1–10m)
Behavior: Sergeant majors get their name from their telltale black bars that resemble military stripes. They are usually found in shallow water, typically along the tops of reefs, and often form large feeding schools of up to a few hundred individuals.
Predators: Grouper and jacks

THREESPOT DAMSELFISH
STEGASTES PLANIFRONS

16

Maximum size: 5in (13cm)
Longevity: Around 15 years
Typical depth: 3–100ft (1–30m)
Behavior: Threespot damselfish tend small gardens of algae. Males use these gardens to attract a mate. If successful, the female will lay her eggs in the male's territory and he will defend them aggressively until they hatch. Threespot damselfish feed on tiny plant-like organisms called epiphytes that grow on the algae they cultivate.
Predators: Grouper and jacks

YELLOWHEAD JAWFISH
OPISTOGNATHUS AURIFRONS

17

Maximum size: 4in (10cm)
Longevity: Unknown, possibly up to 5 years
Typical depth: 10–131ft (3–40m)
Behavior: Jawfish live in burrows in the sediment that they line with stones and bits of crushed shell and coral. Active during the day, they often hover over their burrow and feed on zooplankton. They rarely move far from their burrow and often retreat tail-first when threatened.
Predators: Snapper, grouper and lionfish

ROSY RAZORFISH
XYRICHTYS MARTINICENSIS

18

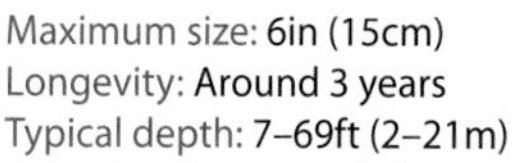

Maximum size: 6in (15cm)
Longevity: Around 3 years
Typical depth: 7–69ft (2–21m)
Behavior: Rosy razorfish are commonly found in open sandy areas near coral reefs. They are active during the day as they feed on small sand-dwelling invertebrates such as crabs, shrimp and worms. Large males often defend a harem of females within their territory.
Predators: Grouper, snapper, barracuda and dolphins

BLUEHEAD WRASSE
THALASSOMA BIFASCIATUM

19

Maximum size: 10in (25cm)
Longevity: 3 years
Typical depth: 0–131ft (0–40m)

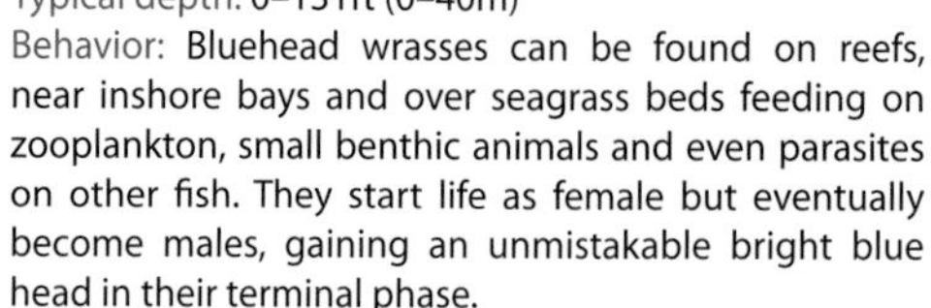

Behavior: Bluehead wrasses can be found on reefs, near inshore bays and over seagrass beds feeding on zooplankton, small benthic animals and even parasites on other fish. They start life as female but eventually become males, gaining an unmistakable bright blue head in their terminal phase.
Predators: Grouper, trumpetfish and soapfish

YELLOWHEAD WRASSE
HALICHOERES GARNOTI

20

Maximum size: 7in (19cm)
Longevity: Unknown, possibly between 3 and 5 years
Typical depth: 3–100ft (1–30m)
Behavior: Yellowhead wrasses are mainly found near coral reefs and rocky ledges. Adults feed on invertebrates while juveniles sometimes clean parasites off larger fish. Yellowhead wrasses are protogynous hermaphrodites, meaning they start life as female but become males at around 3in (7cm) in size.
Predators: Mackerel, grouper and snapper

CREOLE WRASSE
CLEPTICUS PARRAE

21

Maximum size: 12in (30cm)
Longevity: Unknown, but probably around 10 years
Typical depth: 26–328ft (8–100m)
Behavior: Creole wrasses are often found schooling in large numbers above bank reefs, wrecks and on the seaward slopes of reefs. They are most active during the day, when feeding on plankton and small jellyfish. At night, they retreat into reef crevices to sleep.
Predators: Moray eels, grouper and barracuda

SPANISH HOGFISH
BODIANUS RUFUS

22

Maximum size: 16in (40cm)
Longevity: Unknown
Typical depth: 10–230ft (3–70m)
Behavior: Adult Spanish hogfish feed on bottom-dwelling invertebrates, such as brittlestars, crustaceans and sea urchins. Juveniles set up cleaning stations to pick parasites off larger fish. Male hogfish (who start out life as a female) typically manage a harem of three to 12 smaller females.
Predators: Sharks, mackerel and snapper

HOGFISH
LACHNOLAIMUS MAXIMUS

23

Maximum size: 36in (91cm)
Longevity: Up to16 years
Typical depth: 10–100ft (3–30m)
Behavior: Hogfish live in small groups with a dominant male and several smaller females – a common pattern in wrasses. Their name comes from how they root around in the sand with their snout looking for crustaceans and molluscs. Larger individuals frequent the main reef, while smaller individuals are often on patch reefs.
Predators: Sharks and large grouper

PRINCESS PARROTFISH
SCARUS TAENIOPTERUS

24

Max size: 14in (35cm)
Longevity: Unknown, but probably less than 5 years
Typical depth: 3–82ft (1–25m)
Behavior: Princess parrotfish form large schools during the day to feed on plants, algae, sponges and seagrass. Juveniles are more closely associated with seagrass beds. They start out life as female, but can transition to male if no other large breeding males are around.
Predators: Sharks, grouper, jacks and moray eels

QUEEN PARROTFISH 25
SCARUS VETULA

Maximum size: 24 in (61 cm)
Longevity: Up to 20 years
Typical depth: 10–80 ft (3–25 m)
Behavior: During the day, Queen parrotfish feed by scraping algae off rocks and dead coral using their tough, parrot-like beak. At night, Queen parrotfish secrete a membrane of mucus from a gland at the base of the gills which surrounds them like a bubble and masks their scent from nocturnal predators.
Predators: Grouper, eels and sharks

STOPLIGHT PARROTFISH 26
SPARISOMA VIRIDE

Maximum size: 25in (64cm)
Longevity: Around 9 years
Typical depth: 3–164ft (1–50m)
Behavior: Stoplight parrotfish are only active during the day. Their strong beak-like jaws scrape soft algae off the hard coral. They ingest some coral in the process, grinding it up with the help of specialized teeth in their throats and excreting it as coral sand.
Predators: Sharks, barracuda, grouper, snapper, jacks and moray eels

BLUE PARROTFISH 27
SCARUS COERULEUS

Maximum size: 4ft (1.2m)
Longevity: About 10 years
Typical depth: 10–82ft (3–25m)
Behavior: The blue parrotfish is easily identified by its prominent bulging snout and color, as it is the only blue parrotfish. It feeds during the day by biting off pieces of the reef in order to consume plants, algae and small organisms. This species can form large schools, particularly during spawning.
Predators: Large grouper, snapper, moray eels and barracuda

MIDNIGHT PARROTFISH 28
SCARUS COELESTINUS

Maximum size: 30in (77cm)
Longevity: Unknown, but possibly up to 10 years
Typical depth: 16–264ft (5–75m)
Behavior: Midnight parrotfish are among the larger parrotfish species in the Caribbean, and are recognizable for their dark blue-black coloration. They can often be spotted schooling with surgeonfish as they munch on algae-encrusted coral. They are typically associated with coral reefs and sport the telltale beak of all parrotfish.
Predators: Sharks, mackerel and jacks

RAINBOW PARROTFISH

29

SCARUS GUACAMAIA

Maximum size: 4ft (1.2m)
Longevity: Around 10 years
Typical depth: 10–82ft (3–25m)
Behavior: The rainbow parrotfish is the largest herbivorous reef fish in the Caribbean. During the day, it feeds by biting off pieces of the reef in order to consume the plants, algae and small organisms contained within. Schooling may occur in areas where density is high.
Predators: Large grouper, snapper, moray eels, sharks and barracuda

BLACK MARGATE

30

ANISOTREMUS SURINAMENSIS

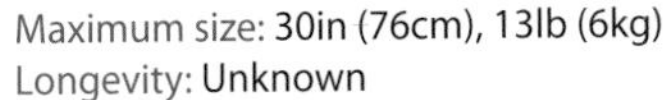

Maximum size: 30in (76cm), 13lb (6kg)
Longevity: Unknown
Typical depth: 0–66ft (0–20m)
Behavior: Black margates can be found alone or in small groups. They are nocturnal, feeding on benthic crustaceans, molluscs, smaller fish and long-spined sea urchins at night. During the day, they can often be found under ledges and sheltering in caves or wrecks.
Predators: Sharks

PORKFISH

31

ANISOTREMUS VIRGINICUS

Maximum size: 16in (40.5cm)
Longevity: Unknown
Typical depth: 0–131ft (0–40m)
Behavior: Porkfish are abundant in Florida waters, particularly on reefs and rocky bottoms. They adapt to new habitats, which makes them common on artificial reefs. They cruise the reef slowly during the day, often in schools. At night, they hunt for molluscs, echinoderms (sea stars and urchins) and crustaceans.
Predators: Grouper and jacks

TOMTATE

32

HAEMULON AUROLINEATUM

Maximum size: 10in (25cm)
Longevity: Unknown, but probably around 10 years
Typical depth: 3–98ft (1–30m)
Behavior: The tomtate is by far the most common member of the grunt family in Florida. This species is found in large schools on coral reefs, but forms pairs for breeding. They have a varied diet consisting of small crustaceans, mollusks, polychaetes, plankton and algae.
Predators: Grouper, snapper, trumpetfish and scorpionfish

BLUE STRIPED GRUNT

HAEMULON SCIURUS

33

Maximum size: 18in (46cm)
Longevity: 12 years
Typical depth: 3–98ft (1–30m)
Behavior: One of the largest members of the grunt family, the blue stripe is also one of the most brightly colored, sporting numerous gold and blue stripes. Juveniles begin life in seagrass beds and move to coral reefs as they become adults. This species can form large schools and is often wary of divers.
Predators: Grouper, snapper, barracuda and sharks

CAESAR GRUNT

HAEMULON CARBONARIUM

34

Max size: 16in (40cm)
Longevity: Unknown, but possibly up to 10 years
Typical depth: 10–82ft (3–25m)
Behavior: Caesar grunts are often found in schools near artificial reefs and over rocky reefs. Like most grunts, they are nocturnal feeders, munching on polychaetes, gastropods and small crustaceans. During the day, they form loose schools under overhangs. Unlike other grunts, however, juveniles settle on shallow reefs and not near mangroves or seagrass beds.
Predators: Sharks, grouper, jacks and moray eels

FRENCH GRUNT

HAEMULON FLAVOLINEATUM

35

Maximum size: 12in (30cm)
Longevity: Unknown, could be up to 12 years
Typical depth: 3–400ft (1–60m)
Behavior: French grunts form large schools on rocky and coral reefs. During the day, adults can often be found resting under ledges and near elkhorn coral. Juveniles spend the day hiding near the shore. French grunts are nocturnal and typically feed on small crustaceans, polychaetes and molluscs.
Predators: Grouper, snapper and trumpetfish

GREY SNAPPER

LUTJANUS GRISEUS

36

Maximum size: 35in (90cm), 44lb (20kg)
Longevity: Around 20 years
Typical depth: 16–590ft (5–180m)
Behavior: Grey snappers are often found schooling, sometimes in large numbers. They feed mainly at night on a range of organisms, including shrimp, crabs, worms and small fishes, rarely moving far to feed.
Predators: Moray eels, sharks, large grouper and barracuda

MAHOGANY SNAPPER
LUTJANUS MAHOGONI

37

Maximum size: 19in (48cm)
Longevity: Around 20 years
Typical depth: 3–330ft (1–100m)
Behavior: This smaller snapper forms large schools during the day, typically in shallower waters over coral reefs. At night, mahogany snapper feed on small fish, shrimp, crabs and cephalopods. They frequent warmer waters and only stray into temperate climates during the heat of summer.
Predators: Sharks, mackerel and other snapper

MUTTON SNAPPER
LUTJANUS ANALIS

38

Maximum size: 37in (94cm), 34lb (15.6kg)
Longevity: 29 years
Typical depth: 82–311ft (25–95m)
Behavior: Mutton snappers can be identified by the small black spot located on their upper back. Many individuals also have one or two blue stripes that run across the cheek and around the eye. Mutton snappers feed both day and night on a mix of fish, crustaceans and gastropods. They are very popular fish with anglers and spearfishers and though size and bag limits exist, the species is still listed as "near threatened" by the IUCN Red List.
Predators: Sharks, large grouper, moray eels, and barracuda

SCHOOLMASTER SNAPPER
LUTJANUS APODUS

39

Maximum size: 26in (67cm)
Longevity: Up to 42 years
Typical depth: 6–207ft (2–63m)
Behavior: Schoolmaster snapper are found in shallow coastal waters in coral reefs and mangrove habitats – adults are often associated with elkhorn corals while younger individuals sometimes enter brackish waters. They feed on crustaceans and cephalopods, although adults also show a preference for fish once their mouth can open wide enough to catch them.
Predators: Sharks, barracuda and grouper

YELLOWTAIL SNAPPER
OCYURUS CHRYSURUS

40

Maximum size: 34in (86cm), 9lb (4kg)
Longevity: Around 13 to 17 years
Typical depth: 3–541ft (1–165m)
Behavior: Yellowtail snapper are typically associated with coral reefs in coastal waters from the U.S. state of Massachusetts, down to the coast of Brazil. They often form schools above reefs and are less commonly seen along the seafloor. They eat plankton and small benthic organisms.
Predators: Sharks, barracuda, mackerel, snapper and grouper

GLASSEYE SNAPPER
HETEROPRIACANTHUS CRUENTATUS

41

Maximum size: 20in (51cm)
Longevity: Unknown
Typical depth: 10–115ft (3–35m)
Behavior: Glasseye snapper are secretive fish, hiding alone or in small groups in holes and crevices during the day. At night, they exit their shelters to feed on octopuses, pelagic shrimp, crabs, small fishes and polychaete worms. They sometimes form larger schools at dusk.
Predators: Sharks, tuna, grouper and mahi mahi

GLASSY SWEEPER
PEMPHERIS SCHOMBURGKII

42

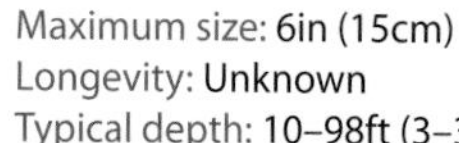

Maximum size: 6in (15cm)
Longevity: Unknown
Typical depth: 10–98ft (3–30m)
Behavior: The nocturnal glassy sweeper is a small fish that feeds on zooplankton and small crustaceans in the water above the reef. During the day, it shelters in groups, hiding in reef crevices and caves. Juveniles are nearly transparent – likely the origin of their name.
Predators: Rays and grouper

BLACKBAR SOLDIERFISH
MYRIPRISTIS JACOBUS

43

Maximum size: 10in (25cm)
Longevity: Unknown
Typical depth: 7–115ft (2–35m)
Behavior: Blackbar soldierfish are nocturnal, often hiding in caves and crevices during the day. They congregate around coral and rocky reefs at night to feed on plankton and invertebrates. They most commonly occur on shallow inshore reefs, but can be found at depths of 330ft (100m).
Predators: Snapper, grouper, jacks and trumpetfish

LONGSPINE SQUIRRELFISH
HOLOCENTRUS RUFUS

44

Maximum size: 14in (35cm)
Longevity: Unknown, but potentially up to 14 years
Typical depth: 0–105ft (0–32m)
Behavior: Longspine squirrelfish often form schools of 8 to 10 individuals at night when they forage for benthic organisms, such as crabs, shrimp, gastropods and brittlestars. During the day, these big-eyed fish seek shelter in holes and crevices in the reef, defending them from other squirrelfish.
Predators: Sharks, grouper, snapper and trumpetfish

CLEANING GOBY 45
ELACATINUS GENIE

Maximum size: 2in (4cm)
Longevity: 3 to 5 years
Typical depth: 3–98ft (1–30m)
Behavior: As their name suggests, cleaning gobies clean other reef creatures by removing their parasites. This behavior is a form of symbiosis known as mutualism, where both parties benefit. The client fish get rid of their ectoparasites, while the cleaners get an easy meal.
Predators: Grouper, snapper and moray eels

REDLIP BLENNY 46
OPHIOBLENNIUS ATLANTICUS

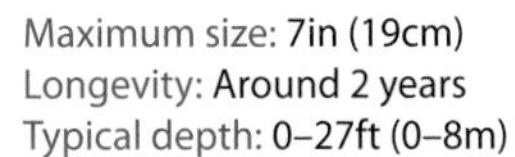

Maximum size: 7in (19cm)
Longevity: Around 2 years
Typical depth: 0–27ft (0–8m)
Behavior: Redlip blennies are common in shallow reef areas with relatively high wave action. Their body shape and modified fins let them "hold on" to the reef. They are herbivorous and territorial, defending a patch of algae during the day and hiding in the reef at night.
Predators: Grouper, snapper and trumpetfish

SHEEPSHEAD PORGY 47
ARCHOSARGUS PROBATOCEPHALUS

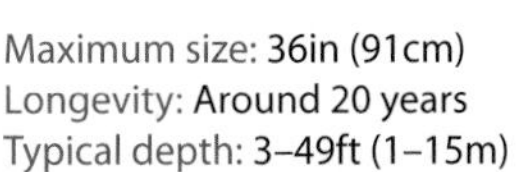

Maximum size: 36in (91cm)
Longevity: Around 20 years
Typical depth: 3–49ft (1–15m)
Behavior: The sheepshead porgy is an important species for the recreational fishing industry in Florida. Also called convict fish due to their striped appearance, they are found in many habitats, including rocky reefs and seagrass beds, as well as around jetties and piers. They are omnivorous, feeding on algae and invertebrates.
Predators: Sharks, grouper and other large carnivorous fish

SPOTTED GOATFISH 48
PSEUDUPENEUS MACULATUS

Maximum size: 12in (30cm)
Longevity: At least 7 years
Typical depth: 0–115ft (0–35m)
Behavior: Spotted goatfish are most often encountered in shallow water over rocky or sandy habitat near reefs. They feed on bottom-dwelling crabs, shrimp and small fish. Spotted goatfish are easily recognizable by the three dark blotches along their back and the telltale barbels they use to stir up the sand when they hunt.
Predators: Sharks, snapper and jacks

YELLOW GOATFISH
MULLOIDICHTHYS MARTINICUS

49

Maximum size: 15in (39cm)
Longevity: Unknown
Typical depth: 0–115ft (0–35m)
Behavior: Yellow goatfish are commonly found swimming in large schools over sandy bottoms. They use their long, sensitive barbels to locate polychaete worms, clams, isopods, amphipods and other crustaceans in the sand. When not feeding, they are often found in groups, sheltering in the reef.
Predators: Sharks, tuna, mahi mahi, grouper and jacks

FLYING GURNARD
DACTYLOPTERUS VOLITANS

50

Maximum size: 21in (50cm)
Longevity: Unknown, but likely more than 5 years
Typical depth: 3–262ft (1–80m)
Behavior: Flying gurnards are often found along sand-bottomed areas near the reef, foraging for benthic crustaceans, crabs, clams and small fishes. They get their name from their fan-like pectoral fins that make it look like they are flying when they swim.
Predators: Sharks, tuna, mahi mahi, grouper and bigeye

LONGLURE FROGFISH
ANTENNARIUS MULTIOCELLATUS

51

Maximum size: 8in (20cm)
Longevity: Unknown, but probably around 10 years
Typical depth: 0–215ft (0–66m)
Behavior: The longlure frogfish is a bottom-dwelling fish that can change color and texture to blend in with its surroundings. It is an ambush predator, feeding mainly on other fish and crustaceans. Frogfish have one of the fastest attacks in the animal kingdom.
Predators: Moray eels and other frogfish

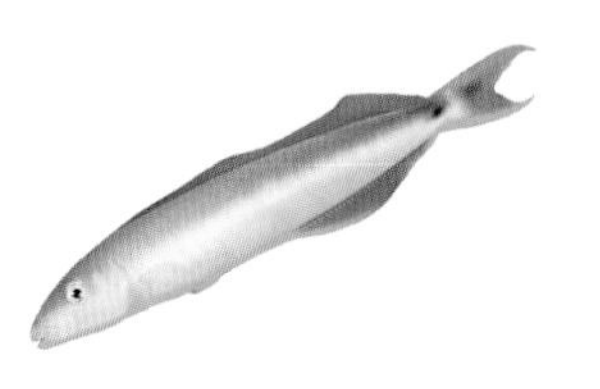

SAND TILEFISH
MALACANTHUS PLUMIERI

52

Maximum size: 28in (70cm)
Longevity: Unknown, but potentially as much as 40 years
Typical depth: 33–164ft (10–50m)
Behavior: Sand tilefish build tunnels and mounds in the sand- and rubble-bottomed areas near reefs and seagrass beds. Tunnel entrances can reach 10ft (3m) in diameter and the mounds are built out of the sand, coral rubble and shell fragments found during excavation.
Predators: Sharks and snapper

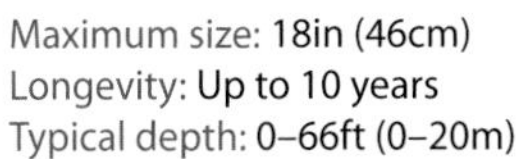

PEACOCK FLOUNDER
BOTHUS LUNATUS

53

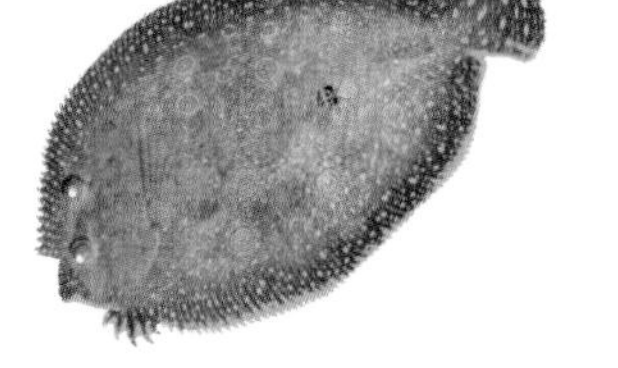

Maximum size: 18in (46cm)
Longevity: Up to 10 years
Typical depth: 0–66ft (0–20m)
Behavior: Peacock flounders are usually found partially buried in loose sand near coral reefs, mangroves and seagrass beds. They are the most common flounders around coral reefs, and mainly feed on small fishes, but also crustaceans and small octopuses.
Predators: Sharks and snapper

BROWN GARDEN EEL
HETEROCONGER LONGISSIMUS

54

Maximum size: 20in (50cm)
Longevity: Unknown
Typical depth: 33–197ft (10–60m)
Behavior: Brown garden eels live in burrows in sandy areas near coral reefs, feeding mainly on plankton that drift by in the current. Individuals rarely leave the safety of their burrows, remaining partially buried with their heads poking out from the sand. They retreat backwards into their burrow when threatened.
Predators: Snake eels and triggerfish

GOLDENTAIL MORAY EEL
GYMNOTHORAX MILIARIS

55

Maximum size: 28in (70cm)
Longevity: Unknown
Typical depth: 0–115ft (0–35m)
Behavior: Goldentail moray eels are common in the Caribbean, living alone in holes and crevices in coral reefs. Unlike other types of morays, goldentails are most active during the day. They feed on fish, molluscs and crustaceans.
Predators: Grouper

SPOTTED MORAY EEL
GYMNOTHORAX MORINGA

56

Maximum size: 3.3ft (1m), 6lbs (2.5kg)
Longevity: Around 10 years, but possibly up to 30 years
Typical depth: 0–656ft (0–200m)
Behavior: Spotted moray eels are most active at night, when they hunt for a wide variety of prey, including parrotfish, grunts, trumpetfish, crustaceans and molluscs. During the day, they are often seen with their head sticking out of a reef hole or crevice.
Predators: Dog snapper and Nassau grouper

JACK-KNIFEFISH
EQUETUS LANCEOLATUS

57

Maximum size: 10in (25cm)
Longevity: Unknown, but possibly as little as 5 years
Typical depth: 33–197ft (10–60m)
Behavior: This member of the drum family is typically found over sandy or muddy bottoms near reefs. It feeds mostly on small bottom-dwelling worms, crustaceans and even organic detritus. The jack-knifefish is highly recognizable by its long, tapered dorsal fin, and its three dark bands. Its striking look has made it valuable in the aquarium trade.
Predators: Sharks, eagle rays and large carnivorous fish

SPOTTED DRUM
EQUETUS PUNCTATUS

58

Maximum size: 11in (27cm)
Longevity: Unknown, but probably around 10 years
Typical depth: 10–98ft (3–30m)
Behavior: Spotted drums are found under ledges, jetties and near small caves during the day. They are solitary and mostly active at night, when they hunt for crabs, shrimp and worms. Drums can emit a drumming sound when they feel threatened, which is the origin of their name.
Predators: Moray eels, grouper and barracuda

SCRAWLED FILEFISH
ALUTERUS SCRIPTUS

59

Maximum size: 43in (110cm), 5.5lbs (2.5kg)
Longevity: Unknown
Typical depth: 10–394ft (3–120m)
Behavior: Scrawled filefish are commonly found on off shore reefs. They are active during the day, feeding on algae, seagrass, hydrozoans, soft corals and anemones. Juveniles sometimes drift with sargassum mats, explaining how this species is found throughout the tropics, and on many non-tropical reefs.
Predators: Barracuda, mahi mahi and large tuna

SLENDER FILEFISH
MONACANTHUS TUCKERI

60

Maximum size: 4in (10cm)
Longevity: Unknown
Typical depth: 7–165ft (2–50m)
Behavior: Slender filefish are almost always found hiding among the branches of gorgonians. During the day, they feed on worms, crabs and zooplankton. At night, they wedge themselves into soft coral with their dorsal spine and stomach appendage, sometimes biting down on a coral polyp while sleeping.
Predators: Grouper and barracuda

OCEAN TRIGGERFISH
CANTHIDERMIS SUFFLAMEN
61

Maximum Size: 25.5in (65cm)
Longevity: Unknown, but probably at least 10 years
Typical depth: 16–98ft (5–30m)
Behavior: Ocean triggerfish spend most of their lives out on the open ocean, hiding from predators among the drifting sargassum and artificial debris. They also visit outer reefs and drop-offs where they form pairs, build nests and reproduce. Ocean triggerfish are found all over the world, from the Caribbean to the Indian and Pacific Oceans.
Predators: Sharks

ATLANTIC TRUMPETFISH
AULOSTOMUS MACULATUS
62

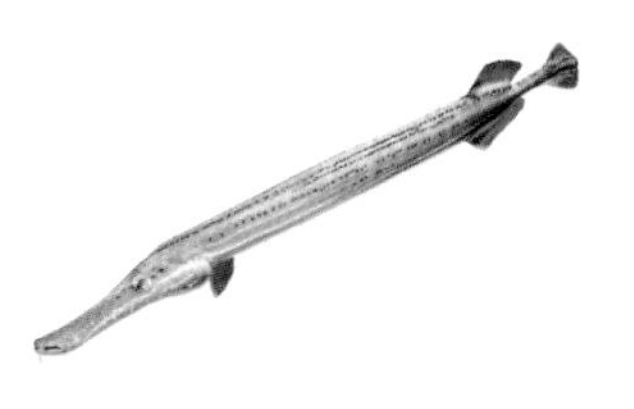

Maximum size: 3ft (1m)
Longevity: Unknown, but likely around 10 years
Typical depth: 6–82ft (2–25m)
Behavior: Trumpetfish are often found camouflaged within the branches of gorgonian corals. They are generally ambush predators that consume small or juvenile reef fish and crustaceans. As known shadow-feeders, they sometimes stalk their prey while swimming alongside other reef fish, using them as cover.
Predators: Grouper, snapper and moray eels

HONEYCOMB COWFISH
ACANTHOSTRACION POLYGONIUS
63

Maximum size: 20in (50cm)
Longevity: Unknown
Typical depth: 7–262ft (2–80m)
Behavior: Honeycomb cowfish are protected by hexagon scales that form a rigid carapace over much of their bodies. They are relatively slow and wary, which makes their external armor an essential defense against potential predators. They usually forage alone, feeding on sponges, tunicates and shrimp.
Predators: Sharks

SMOOTH TRUNKFISH
LACTOPHRYS TRIQUETER
64

Maximum size: 18.5in (47cm)
Longevity: Unknown
Typical depth: 0–164ft (0–50m)
Behavior: Smooth trunkfish are easily recognized by their black mouth and white-spotted, triangular, armored shape. They are not fast swimmers, relying instead on their armor and toxins to deter predators. They are easily approached and can often be seen hunting bottom invertebrates by jetting water from their mouth to disturb the sand and locate their prey.
Predators: Mahi mahi, cobia and large carnivorous fish

PORCUPINEFISH
DIODON HYSTRIX

65

Maximum size: 35in (90cm)
Longevity: Up to 10 years
Typical depth: 7–164ft (2–50m)
Behavior: Porcupinefish are solitary nocturnal predators that feed on snails, crabs and sea urchins. During the day, they are often found sheltering in reef caves or crevices. They can inflate their bodies up to twice their normal size by drawing in water or air.
Predators: Dolphins, and large pelagic fish such as sharks and tuna

CARIBBEAN SHARPNOSE PUFFER
CANTHIGASTER ROSTRATA

66

Maximum size: 5in (12cm)
Longevity: Unknown, but possibly up to 10 years
Typical depth: 3–130ft (1–40m)
Behavior: Sharpnose puffers prefer reefs where gorgonian corals are common. They are most active during the day as they search for small reef invertebrates such as crabs, shrimp, worms and snails. They are territorial, so if you happen to see two individuals near one another, they may be engaged in defensive displays.
Predators: Grouper, snapper, barracuda and moray eels

CONEY
CEPHALOPHOLIS FULVA

67

Maximum size: 17in (43cm)
Longevity: 11 years, possibly as much as 19 years
Typical depth: 3–148ft (1–45m)
Behavior: Coney hide in caves and crevices in the reef during the day, venturing out at night to forage for small reef fish and crustaceans. They are approachable but wary, and the males are territorial. They start out as female, becoming male at around 8in (20cm).
Predators: Sharks, grouper and snapper

GRAYSBY
CEPHALOPHOLIS CRUENTATA

68

Maximum size: 16in (40 cm)
Longevity: Approximately 12 years
Typical depth: 7–561ft (2–170m)
Behavior: Graysbies are found in reef areas that contain caves, crevices or hollow sponges where they hide during the day. At night, they hunt for reef fish, such as chromis, squirrelfish, gobies and crustaceans. Some individuals hunt alongside moray eels at night.
Predators: Barracuda, sharks and larger grouper

BLACK GROUPER

MYCTEROPERCA BONACI

69

Maximum size: 5ft (1.5m), 220lbs (100kg)
Longevity: More than 30 years
Typical depth: 19–246ft (6–75m)
Behavior: Black grouper are abundant in Florida waters but rarely seen. They tend to shy away from swimmers. Commercially fished in many places, their populations are generally declining as a result. They are solitary except when they congregate to spawn. Adults feed on smaller reef fish such as grunts and snapper.
Predators: Sharks

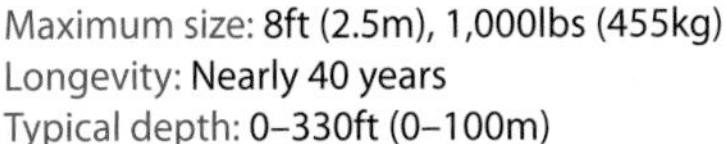

GOLIATH GROUPER

EPINEPHELUS ITAJARA

70

Maximum size: 8ft (2.5m), 1,000lbs (455kg)
Longevity: Nearly 40 years
Typical depth: 0–330ft (0–100m)
Behavior: Goliath grouper are the largest grouper species in Florida. This massive, solitary fish does not have a large home range, but will defend its territory aggressively against intruders by making loud "barking" noises with its swim bladder. They have even been known to charge divers, so beware. Goliath grouper feed on lobsters, fish and even turtles and stingrays.
Predators: Sharks

BUTTER HAMLET

HYPOPECTRUS UNICOLOR

71

Maximum size: 5in (13cm)
Longevity: Unknown
Typical depth: 23–82ft (7–25m)
Behavior: There is some scientific debate as to whether the various hamlet species are in fact different species at all. They are typically differentiated by their coloration, and there is little genetic difference among species. Butter hamlets are associated with reefs and feed on small reef fish and benthic crustaceans.
Predators: Grouper, snapper, jacks and barracuda

FAIRY BASSLET

GRAMMA LORETO

72

Maximum size: 3in (8cm)
Longevity: Around 6 years, but up to 12 years
Typical depth: 3–180ft (1–60m)
Behavior: Fairy basslets are often found on reef walls that are full of caves and ledges. They are most active during the day, feeding mainly on crustaceans – although they occasionally act as a cleaner fish. At night, they retreat into the safety of a familiar reef shelter.
Predators: Snapper, grouper and moray eels

TOBACCOFISH
SERRANUS TABACARIUS

73

Maximum size: 7in (18cm)
Longevity: Up to 20 years
Typical depth: 10–230ft (3–70m)
Behavior: Tobaccofish are found in a range of habitats, from sand and rubble areas to reef walls. They are most active during the day, when they hunt for small fish and crustaceans, sometimes alongside goatfish. At night, they hide in reef cracks and crevices.
Predators: Grouper, jacks and eels

ATLANTIC SPADEFISH
CHAETODIPTERUS FABER

74

Maximum size: 35in (90cm)
Longevity: Up to 20 years
Typical depth: 10–115ft (3–35m)
Behavior: Atlantic spadefish are often found in schools of up to 500 individuals, swimming above reefs and shipwrecks. They feed during the day on plankton and benthic invertebrates, such as worms, crustaceans and molluscs. To hide from predators, juveniles often drift on their side to mimic debris.
Predators: Grouper and sharks

BERMUDA CHUB
KYPHOSUS SECTATRIX

75

Maximum size: 30in (76cm), 13lbs (6kg)
Longevity: Unknown
Typical depth: 3–330ft (1–10m)
Behavior: Bermuda chub are a schooling fish found in shallow waters above sandy areas and seagrass beds, and near coral reefs. They feed on benthic algae, but also on small crabs and molluscs. Juveniles often associate with floating sargassum mats, letting them disperse across great distances.
Predators: Sharks, barracuda, snapper, moray eels and scorpionfish

BAR JACK
CARANX RUBER

76

Maximum size: 23in (59cm)
Longevity: Unknown, possibly up to 30 years
Typical depth: 3–330ft (1–100m)
Behavior: Bar jacks sometimes swim alone, but are usually found schooling in shallow, clear water near coral reefs. They feed on fish, shrimp and other invertebrates. They are the most abundant species of jack in the Caribbean, and are easily approached by divers.
Predators: Grouper, mackerel, mahi mahi and large jacks

HORSE-EYE JACK
CARANX LATUS

77

Maximum size: 3ft (1m), 29lbs (13kg)
Longevity: Unknown
Typical depth: 3–66ft (0–20m)
Behavior: Horse-eye jacks are schooling pelagic fish that frequent the waters above off shore reefs, although juveniles are often seen inshore along sandy beaches. Adults feed on fish, shrimp and other invertebrates. They often approach divers boldly, but without posing much of a threat.
Predators: Sharks, barracuda and mahi mahi

PALOMETA / GREAT POMPANO
TRACHINOTUS GOODEI

78

Maximum size: 20in (50cm)
Longevity: Unknown
Typical depth: 0–40ft (0–12m)
Behavior: Adult palometa tend to form schools over shallow coral reefs. Juveniles, meanwhile, are more common over sand and rubble habitat. They are most active during the day, feeding on crustaceans, worms, molluscs and fish.
Predators: Sharks and barracuda

CERO
SCOMBEROMORUS REGALIS

79

Maximum size: 6ft (1.8m)
Longevity: Unknown
Typical depth: 3–66ft (1–20m)
Behavior: A member of the mackerel family, the cero is typically found swimming in open water near coral reefs, occasionally in schools. Ceros eat smaller fish, including herring, anchovies and silversides, along with squid and shrimp. They are considered a good game fish.
Predators: Sharks, tuna, marlin, king mackerel and wahoo

TARPON
MEGALOPS ATLANTICUS

80

Maximum size: 8ft (2.5m), 330lb (150kg)
Longevity: Around 50 years
Typical depth: 3–330ft (0–100m)
Behavior: Tarpon frequent both marine and freshwater ecosystems, from Canada in the north to Brazil in the south. They can feed during both the day and the night on a range of fish and crustaceans. Tarpon have relatively small teeth and tend to swallow their prey whole.
Predators: Sharks and dolphins

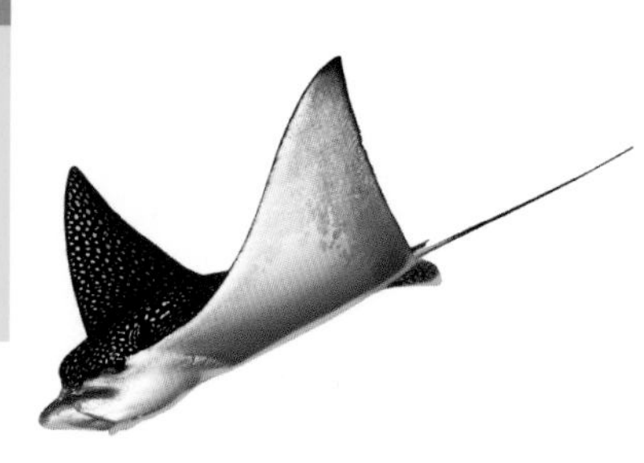

SPOTTED EAGLE RAY
AETOBATUS NARINARI

81

Maximum size: 10ft (3m) disc width, 500lb (230kg)
Longevity: Up to 20 years
Typical depth: 3–260ft (1–80m)
Behavior: Spotted eagle rays are carnivores that specialize in eating hard-shelled prey such as conch, clams, crabs and lobsters. They sometimes eat octopuses and fish as well, and are often found over sand habitat. They have electro-receptors in their snout to help search for buried prey.
Predators: Tiger, bull, lemon and hammerhead sharks

NURSE SHARK
GINGLYMOSTOMA CIRRATUM

82

Maximum size: 14ft (4.3m), 242lbs (110kg)
Longevity: Up to 25 years
Typical depth: 0–430ft (0–130 m)
Behavior: Nurse sharks are large nocturnal reef predators. At night, they search for hard-shelled prey, such as lobsters, crabs and conch, which they consume with their specially designed jaws. During the day, they are often found resting in caves or beneath coral overhangs.
Predators: Larger shark species

LONGSNOUT SEAHORSE
HIPPOCAMPUS REIDI

83

Maximum size: 7in (18cm) with tail outstretched
Longevity: Unknown, but probably at least 4 to 5 years
Typical depth: 3–55m (10–180ft)
Behavior: Seahorses are rare throughout the Caribbean. They prefer shallow reef areas. They are often seen clinging to seagrass, macroalgae, gorgonians and sponges with their prehensile tails, while they feed on zooplankton, mysid shrimp and small crustaceans.
Predators: Rays, turtles and crabs

PEDERSON CLEANER SHRIMP
ANCYLOMENES PEDERSONI

84

Maximum size: 1in (3cm)
Longevity: Unknown
Typical depth: 3–115ft (1–35m)
Behavior: Pederson cleaner shrimp pick parasites off reef fish. They are found in close association with sea anemones, which help advertise the shrimp's cleaning services and provide shelter. The anemone's stinging tentacles ward off predators but do not sting its resident cleaner shrimp, which can number up to a dozen.
Predators: Unknown

BANDED CORAL SHRIMP
STENOPUS HISPIDUS

85

Maximum size: 4in (10cm)
Longevity: Around 3 years
Typical depth: 6–656ft (2–200m)
Behavior: Banded coral shrimp are often found hiding in reef cracks and sponges. They are most active at night when hunting for small fish, other crustaceans, snails and worms, although they sometimes clean parasites from other reef creatures. This species of shrimp forms monogamous pairs that defend a territory.
Predators: Grouper, snapper, moray eels and barracuda

YELLOWLINE ARROW CRAB
STENORHYNCHUS SETICORNIS

86

Maximum size: 2in (6cm)
Longevity: Around 5 years
Typical depth: 10–130ft (3–40m)
Behavior: Yellowline arrow crabs are small spider-like creatures with triangular bodies and small purple claws. They are often found inside tube sponges, and among the tentacles of anemones and spines of sea urchins. At night, they forage for algae, detritus, tube worms and bristleworms.
Predators: Grouper, puffers, triggerfish, wrasses and grunts

CARIBBEAN SPINY LOBSTER
PANULIRUS ARGUS

87

Maximum size: 18in (45cm)
Longevity: Around 20 years
Typical depth: 0–295ft (0–90m)
Behavior: Caribbean spiny lobsters like to hide in reef caves and crevices during the day. At night, they roam the reef searching for snails, clams, crabs and dead and decaying organisms to eat. They undergo seasonal mass migrations in the fall, marching in single-file towards deeper water.
Predators: Sharks, stingrays, grouper, triggerfish and moray eels

CARIBBEAN REEF SQUID
SEPIOTEUTHIS SEPIOIDEA

88

Maximum size: 8in (20cm)
Longevity: Around 1 year
Typical depth: 0–98ft (0–30m)
Behavior: Caribbean reef squid are often found in small schools. They capture food in their 10 arms, feeding mainly on small fishes, as well as crustaceans and other molluscs. They have the largest eyes relative to body size of any animal, and they track their food by sight.
Predators: Grouper, snapper and barracuda

FLAMINGO TONGUE
CYPHOMA GIBBOSUM

89

Maximum size: 2in (4cm)
Longevity: Unknown, but likely 2 years
Typical depth: 6–45ft (2–14m)
Behavior: Flamingo tongues are a reef gastropod (marine snail) almost always found feeding on sea fans, sea whips and other gorgonians. The flesh of the gorgonians they eat contains toxic chemicals that the flamingo tongue converts into its own predator-deterring toxins.
Predators: Hogfish

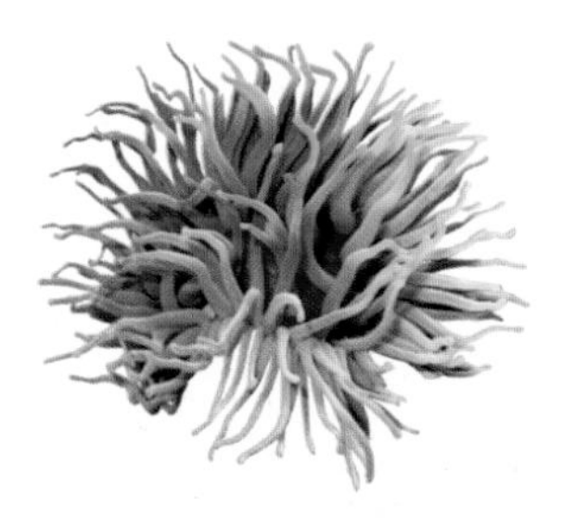

GIANT ANEMONE
CONDYLACTIS GIGANTEA

90

Maximum size: 12in (30cm)
Longevity: Around 75 years
Typical depth: 3–82ft (1–25m)
Behavior: Giant anemones come in a variety of colors, from white to dark brown, sometimes with pink or purple-tipped tentacles. Some individuals have stinging tentacles that they use to paralyze and capture fish, shrimp and worms. Anemones typically attach to the reef, but are also capable of crawling.
Predators: Hermit crabs, snails and sea slugs

YELLOW TUBE SPONGE
APLYSINA FISTULARIS

91

Maximum size: 3.3ft (1m)
Longevity: Unknown, but likely up to hundreds of years
Typical depth: 10–246ft (3–75m)
Behavior: The yellow tube sponge is a filter feeder that draws water into its structure in order to filter out food, such as plankton, detritus and bacteria suspended in the water. They do this continuously, night and day. Sponges are attached to the reef and unable to move. However, if a sponge is broken into pieces, each piece can grow into a new sponge.
Predators: Hawksbill sea turtles

GIANT BARREL SPONGE
XESTOSPONGIA MUTA

92

Maximum size: 3ft (1m) in height, 6ft (1.8m) across
Longevity: Up to 2,000 years, perhaps longer
Typical depth: 33–99ft (10–30m)
Behavior: Giant barrel sponges are often called the redwoods of the coral reef due to their massive size. As a group, sponges have been around for 500 million years. They draw water through their walls, filter out food particles, then expel the waste water through the osculum opening at the top. They provide shelter for countless reef fish and invertebrates.
Predators: Green and hawksbill sea turtles, sea slugs, crabs, parrotfish, rock beauties and many other reef creatures

Ian POPPLE
ian.popple@reef-smart.com

Born and raised in the U.K., Ian earned his undergraduate degree in Oceanography from the University of Plymouth in 1994. He worked for five years at Bellairs Research Institute in Barbados, supporting research projects across the region, before completing his Master's in marine biology at McGill University in 2004. He co-founded a marine biology education company, Beautiful Oceans, before founding Reef Smart in 2015, to raise awareness and encourage people to explore the underwater world. Ian has published in both the scientific and mainstream media, including National Geographic, Scuba Diver Magazine and the Globe and Mail. He is a PADI Dive Instructor with over 3,000 dives in 30 years of diving experience.

Otto WAGNER
otto.wagner@reef-smart.com

Born and raised in Romania, Otto graduated from the University of Art and Design in Cluj, Romania in 1991. He moved to Canada in 1999 where he studied Film Animation at Concordia University in Montreal. In 2006, Otto turned to underwater cartography and pioneered new techniques in 3D visual mapping. He co-founded Art to Media and began mapping underwater habitats around the world. Throughout his 25-year career, Otto has received numerous awards and international recognition for his work, including the Prize of Excellence in Design from the Salon International du Design de Montréal. He has also illustrated seven books. Otto is a PADI Advanced Diver with over 500 dives in more than 15 years of diving experience.

Peter McDOUGALL
peter.mcdougall@reef-smart.com

Born and raised in Canada, Peter received his undergraduate and Master's degrees from McGill University. His focus on behavioral ecology and coral reef ecology led him to two field seasons at Bellairs Research Institute in Barbados, in 1999 and again in 2002. After graduating in 2003, Peter moved to the United States and began a career in science communication and writing, publishing in both peer-reviewed academic journals and the popular press. He has written on a variety of coastal ecosystem issues, including extensive work surrounding the science of ocean acidification. He is a PADI Rescue Diver with over 300 dives in 20 years of experience.

REEF SMART GUIDES
SCUBA DIVE SNORKEL SURF
FORT LAUDERDALE
POMPANO BEACH AND DEERFIELD BEACH
www.reefsmartguides.com

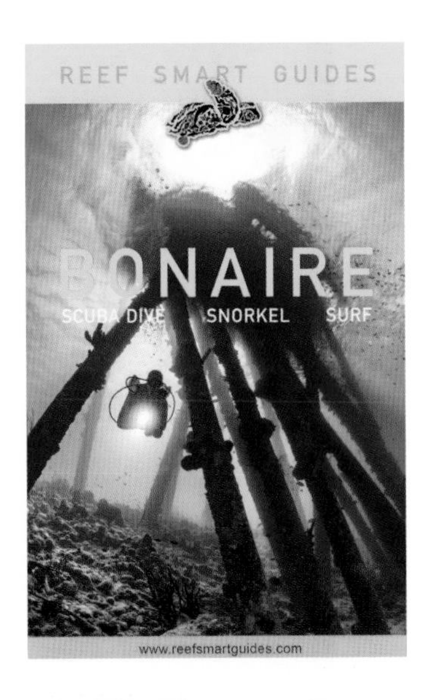
REEF SMART GUIDES
BONAIRE
SCUBA DIVE SNORKEL SURF
www.reefsmartguides.com

REEF SMART GUIDES
BARBADOS
SCUBA DIVE SNORKEL SURF
www.reefsmartguides.com

REEF SMART GUIDES
GRAND CAYMAN
SCUBA DIVE SNORKEL SURF
Coming in 2019

REEF SMART GUIDES
BONAIRE
SCUBA DIVE SNORKEL SURF

REEF SMART GUIDES
BARBADOS
SCUBA DIVE SNORKEL SURF

REEF SMART GUIDES
SCUBA DIVE SNORKEL SURF
FORT LAUDERDALE
POMPANO BEACH AND DEERFIELD BEACH

REEF SMART GUIDES
PALM BEACH
SCUBA DIVE SNORKEL SURF

8:58 PM
SALT PIER
SALT CITY
North side of Salt Pier
188